AF783861

Food Chain Security

NATO Science for Peace and Security Series

This Series presents the results of scientific meetings supported under the NATO Programme: Science for Peace and Security (SPS).

The NATO SPS Programme supports meetings in the following Key Priority areas: (1) Defence Against Terrorism; (2) Countering other Threats to Security and (3) NATO, Partner and Mediterranean Dialogue Country Priorities. The types of meeting supported are generally "Advanced Study Institutes" and "Advanced Research Workshops". The NATO SPS Series collects together the results of these meetings. The meetings are co-organized by scientists from NATO countries and scientists from NATO's "Partner" or "Mediterranean Dialogue" countries. The observations and recommendations made at the meetings, as well as the contents of the volumes in the Series, reflect those of participants and contributors only; they should not necessarily be regarded as reflecting NATO views or policy.

Advanced Study Institutes (ASI) are high-level tutorial courses intended to convey the latest developments in a subject to an advanced-level audience

Advanced Research Workshops (ARW) are expert meetings where an intense but informal exchange of views at the frontiers of a subject aims at identifying directions for future action

Following a transformation of the programme in 2006 the Series has been re-named and re-organised. Recent volumes on topics not related to security, which result from meetings supported under the programme earlier, may be found in the NATO Science Series.

The Series is published by IOS Press, Amsterdam, and Springer, Dordrecht, in conjunction with the NATO Public Diplomacy Division.

Sub-Series

A.	Chemistry and Biology	Springer
B.	Physics and Biophysics	Springer
C.	Environmental Security	Springer
D.	Information and Communication Security	IOS Press
E.	Human and Societal Dynamics	IOS Press

http://www.nato.int/science
http://www.springer.com
http://www.iospress.nl

Series C: Environmental Security

Food Chain Security

edited by

Hami Alpas
Middle East Technical University
Ankara, Turkey

and

Beyazit Cirakoglu
Acibadem University
Istanbul, Turkey

Springer

Published in cooperation with NATO Public Diplomacy Division

Proceedings of the NATO NFA pilot study in a series of meetings on
Food Chain Security
Antalya, Turkey
final meeting: 19–20 September 2008

Library of Congress Control Number: 2010933791

ISBN 978-90-481-9560-2 (PB)
ISBN 978-90-481-9557-2 (HB)
ISBN 978-90-481-9558-9 (eBook)

Published by Springer,
P.O. Box 17, 3300 AA Dordrecht, The Netherlands.

www.springer.com

Printed on acid-free paper

All Rights Reserved
© Springer Science + Business Media B.V. 2010
No part of this work may be reproduced, stored in a retrieval system, or transmitted in any form or by any means, electronic, mechanical, photocopying, microfilming, recording or otherwise, without written permission from the Publisher, with the exception of any material supplied specifically for the purpose of being entered and executed on a computer system, for exclusive use by the purchaser of the work.

Contents

Preface

The pilot study on Food Chain Security was launched in 2003 by NATO Public Diplomacy Division Science for Peace and Security Section (SPS) under the leadership of Turkey.

The purpose of the study was to study the safety and security of food stuffs in the face of their careless/ignorant handling as well as against expected terrorist attacks at the system which may destroy and/or degrade it at the source during distribution, processing and in the consumption phase. The study included the protective and response measures which may have to be taken to reduce the risk and mitigate the consequences of these threats to the food system. The final outputs of this pilot study were agreed to be mainly:

To allow comparison between country partners
To identify common weaknesses of the food systems

As a result of the terrible September 11, 2001 attacks in the United States the nature of the terrorist threat appears to be more uncertain and diffused, therefore the terrorist threat against the food system which comprises production, processing, distribution, restaurants, and retail can be very diverse and unpredictable and involve chemical, biological, and radiological agents of various kinds. Preparing for all possible contingencies was not practical, so a "risk management approach" was used in this study based on risk management principles that acknowledge while risk generally cannot be eliminated, enhancing protection from known or potential threats can reduce it.

During the 5 year period from October 2003 to September 2008, nine meetings were conducted: eight Pilot Study meetings and one Steering Committee meeting (Brussels, May 2008). Also one topical workshop on "Food Safety in Russian Federation and Eastern Europe, Caucasus and Central Asia (EECCA) Countries" (Moscow, Russian Federation, October 2004) was organized. Therefore this book reports on a success story; which is the efficient collaboration between scientists from Europe, Mediterranean Region, North America and Asia on a topic such as "Food Security" where nobody is "immune from the risks" and concerns all countries and societies in any part of the world in a period of increased instability and limitation of resources.

The issues related with Food Security will continue to be a priority topic for all countries as it is a global issue that is continually developing. The results of this pilot study are relevant for many partner countries that will be facing these issues. Many of the countries in the pilot study expressed a need to create additional fora and workshops to continue the dialogue on this issue among the professionals and efforts are continuing.

We do also hope that enough reference has been made to the audience of this book to create opportunities for further collaboration and related sources for possible funding. As a consequence, we will be very satisfied to see if this book serves as a helpful tool both in becoming more acquainted with the issue of food security, and also in being aware of the possible ways to implement further research and analysis on this important issue.

We would like to take this opportunity to thank to NATO Public Diplomacy Division Science for Peace and Security Section (SPS) for the continuing support throughout all phases of this study; especially to Dr. Deniz Beten – Programme Director Environmental Security, Sue Williamson and Martine Deweer.

Last but not least we would like to dedicate this book to the memory of our distinguished colleague Biljana Miljovska from Macedonia who passed away in 2004.

Ankara, Turkey

H. Alpas
B. Cirakoglu

Contributors

Alpas, Hami
Food Engineering Department, Middle East Technical University
Ankara, 06531, Turkey, imah@metu.edu.tr

Brunet, Sebastien
Dépt. de Science Politique, University of Liege
7, Bd du Rectorat, B 31, Bte 29, 4000 Liege, Belgium,
Sebastien.Brunet@ulg.ac.be

Claisse, Frédéric
Dépt. de Science Politique, University of Liege
7, Bd du Rectorat, B 31, Bte 29, 4000 Liege, Belgium,
Sebastien.Brunet@ulg.ac.be

Delvenne, Pierre
Dépt. de Science Politique, University of Liege
7, Bd du Rectorat, B 31, Bte 29, 4000 Liege, Belgium,
Sebastien.Brunet@ulg.ac.be

Empis, José
Institute for Biotechnology and Bioengineering
CEQB do IST, Av. Rovisco Pais, 1049-001 Lisboa, jempis@ist.utl.pt

Ermakova, Irina V.
Institution of Russian Academy of Sciences
Institute of Higher Nervous Activity and Neurophysiology of RAS
Butlerov str., 5a, Moscow 117485, Russian Federation, ermak_i@mail.ru

Galatchi, Liviu-Daniel
Ovidius University of Constanta
Constanta-4, Romania, galatchi@univ-ovidius.ro

Gaus, Joachim
Federal Ministry of Food, Agriculture and Consumer Protection
Wilhelmstraße 54, 10117 Berlin, Germany, Joachim.Gaus@bmelv.bund.de

Jackson, LeeAnne
Food and Drug Administration
Center for Food Safety and Applied Nutrition
Office of Food Defense, Communication, and Emergency Response
Food Defense Oversight Team, 5100 Paint Branch Parkway HFS-007, College Park, MD 20740, USA, LeeAnne.Jackson@fda.hhs.gov

Komar, Ales
University of Defence, 65 Kounicova, 612 00 Brno
Czech Republic, ales.komar@unob.cz

Mihalache, Diana-Lacramioara
Ovidius University of Constanta
Constanta-4, Romania, liviugalatchi@yahoo.com

Sivonen, Hannu
National Emergency Supply Agency, Pohjoinen Makasiinikatu 7 A
FI-00130 Helsinki, Finland, hannu.sivonen@nesa.fi

Smith, Madeleine
Chemical Engineering, University of Birmingham
Edgbaston, Birmingham, United Kingdom, B15 2TT, m.smith.2@bham.ac.uk

Vasicka, Pavlina
University of Defence, 65 Kounicova, 612 00 Brno
Czech Republic, ales.komar@unob.cz

Veiga, Alexandra
Instituto de Tecnologia Química e Biológica
Universidade Nova de Lisboa, Av. da República
2780-157 Oeiras, Portugal, aveiga@itqb.unl.pt

Food Chain Security Pilot Study: Findings and Recommendations

Hami Alpas

Food Engineering Department, Middle East Technical University, Ankara, 06531, Turkey

1. Introduction and Purpose

The pilot study on Food Chain Security was launched in 2003 by the NATO Committee of the Challenges of Modern Society (CCMS) which was renamed Science for Peace and Security (SPS) under the leadership of Turkey.

The purpose of the study was to study the safety and security of food stuffs in the face of their careless/ignorant handling as well as against expected terrorist attacks at the system which may destroy and/or degrade it at the source during distribution, processing and in the consumption phase. The study included the protective and response measures which may have to be taken to reduce the risk and mitigate the consequences of these threats to the food system.The final outputs of this pilot study were agreed to be mainly:

To allow comparison between country partners;
To identify common weaknesses of the food systems.

As a result of the terrible September 11, 2001 attacks in the United States the nature of the terrorist threat appears to be more uncertain and diffused, therefore the terrorist threat against the food system which comprises production, processing, distribution, restaurants, and retail can be very diverse and unpredictable and involve chemical, biological, and radiological agents of various kinds. Preparing for all possible contingencies was not practical, so a "risk management approach" was used in this study based on risk management principles that acknowledge while risk generally cannot be eliminated, enhancing protection from known or potential threats can reduce it.

2. Activities

During the 5 year period from October, 2003 to September, 2008, nine meetings were conducted: eight Pilot Study meetings and one Steering Committee meeting (Brussels, May 2008). Also one topical workshop on "Food Safety in Russian

H. Alpas and B. Cirakoglu (eds.), *Food Chain Security*,
DOI 10.1007/978-90-481-9558-9_1, © Springer Science + Business Media B.V. 2010

Federation and Eastern Europe, Caucasus and Central Asia (EECCA) Countries" (Moscow, Russian Federation, October 2004) was organized.

The Pilot Study Director, Reporter and group members also participated and contributed to the following activities:

NATO NRC-CCMS Short Term Project on "Ecoterrorism" (Russian Federation-Italy, 2005-2007)

International Seminar on "Securing Food Supplies in Europe– Risk Scenarios "(Vienna, Austria, December 2004)

COST - NATO Strategic Expert Meeting on "Food Security and Simulation" (Brussels, Belgium, November 2006)

First European Food Congress-NATO-SPS: Food Security and Related Items (Ljubljana, Slovenia, November 2008)

NATO-ARW "Threats to Food and Water Chain Infrastructure" (Vienna, Austria, December, 2008).

3. Meetings

3.1. First Pilot Study Meeting: October 2–3, 2003, Istanbul, Turkey

The purpose of the meeting was to determine the most effective manner in which this Pilot Study could address the key issues related to Food Chain Security. In this respect a distinction between Food Safety and Food Security was made. It was also concluded that this Pilot Study should make the rules and should act as "decision makers". An overall agreement on generic models with a framework approach of counter terrorism on food was strongly emphasized for the final success of the pilot study. Using the FAO definition of food security:

"Food security exists when all people at all times have physical and economic access to sufficient, safe and nutritious food to meet their dietary needs and food preferences for an active and healthy life style" (FAO 1996)

It was agreed that safety is at the first plan and security is a response and in this respect the following working groups were formed to be detailed in the second meeting.

Overview of the food system
Surveillance and detection systems
Response system

3.2. Second Pilot Study Meeting: May 21–22, 2004, Liege, Belgium

The main purpose of this meeting was a detailed description of the responsibilities, areas covered and working schemes of the groups formed in the first meeting. After combining all the presentations and remarks, the Pilot Study group agreed to detail and expand the working groups (WG's):

WG 1: Overview of the food system
Regulatory agencies and organizational arrangements
Picture of food sector in each participating country
Nature and the likelihood of the threat
Preventive measures
WG 2: Surveillance and detection systems
Medical
Veterinarian
Agriculture
Commercial
Custom service
Consumer's view
WG 3: Response system
Alert systems
Recall system
Care systems
Emergency measures

The working groups participated to the same general picture of food chain security and followed a logical path from the description of the food system, to the surveillance and response systems. So, in order to be effective, all participants provided each of the working groups with data and/or a case study. The guidelines and questioning for each working group was elaborated by the coordinator and the participants. Overlaps and contradictions were avoided by reviewing the guidelines and questioning by the director and reporter of the pilot study.

In the light of the contributions made during the second Pilot Study meeting a topical workshop on "Food Safety in Russian Federation and Eastern Europe, Caucasus and Central Asia (EECCA) Countries" was jointly conducted by the RF Ministry of Natural Resources and the Russian Federation Regional Ecological Centre, in conjunction with the NATO-CCMS, on October 28–29, 2004, Moscow, Russian Federation. The following were the main outputs-goals of the meeting:

- To continue exchange of experience between participating countries in the area of legal regulations, logistical and institutional aspects of food safety in the NATO-Russia Council (NRC) cooperation framework between the RF Ministry of Natural Resources and NATO-CCMS

- To note the urgency of further study of issues connected with the production and use of genetically modified foodstuffs and feeds, stressing by the same token the complexity of the question and the need to devise common European standards
- To explore the possibility of organizing a pilot project involving Russian Federation and EECCA experts under the umbrella of NATO-CCMS food safety research
- To recommend to participating countries that they involve non-governmental and voluntary organizations in decision-making and monitoring on food-safety problems

3.3. Third Pilot Study Meeting: March 10–11, 2005, Antalya, Turkey

The output of the workshop conducted in Moscow, Russian Federation was discussed in detail and in the light of the agreed goals the Pilot Study group agreed to continue the exchange of experience between participating countries in the area of legal regulations, logistical and institutional aspects of food safety in the NRC cooperation framework between the RF Ministry of Natural Resources and NATO-CCMS through the ongoing Pilot Study on "Food Chain Security". In this respect the Italian–Russian Federation proposal on NATO NRC-CCMS Short Term Project on "Ecoterrorism" (Russian Federation-Italy, 2005–2007) was supported with participation of experts from our Pilot Study Group.

The activities of the three working groups (WG) were summarized. In this respect in order to fill gaps for certain types of information it was necessary to gather information by using a questionnaire-survey. The survey was shared in order to show to the target audience "what could be the potential effects if they are not implementing the guidelines".

The main objectives of the survey (http://www.nato.int/science/pilot-studies/fcs/060116_report_washington.pdf) were:

To identify the concerns of the food industries with regard to contamination of the food chain (i.e. what did the industry perceive as the main hazards associated with the food chain)

To evaluate the extent to which food safety management systems had been implemented as a safeguard against contamination incidents

To determine the feasibility and limitations of conducting such a survey within participating countries

3.4. Fourth Pilot Study Meeting: November 1–2, 2005, Alexandria, Virginia, USA

The preliminary outputs of the survey data were discussed to be more concrete in terms of "risk mapping" mainly from participating countries (Bulgaria, Germany, Portugal, Spain and United Kingdom).

It was agreed that the information flow from bottom to top in a time of crisis does work properly in the light of the experience(s) of EU-member states with "Rapid Alert system for Food and Feed (RASFF)". The two EU regulations (EC No 178/2002) and (EC No 882/2004) was discussed in terms of similarities and differences in terms of both the applications within EU-member states and other participating countries of the Pilot Study which are not EU-members. The technical discussions were also conducted by experts from the Centers for Disease Control and Prevention (CDC-USA), Environmental Protection Agency (EPA-USA), and the Food and Drug Administration (FDA-USA). Although there were differences in applications in participating countries; each country participating in the pilot study had response protocols for crises. However it was concluded that traditional physical security practices alone cannot protect the sector as agriculture and food systems are extensive, open, interconnected, diverse, and complex structures providing attractive potential targets for terrorist attacks. Due to the rapidity by which food products move in commerce to consumers and the time required for detection and identification of a causative agent, attacks on the food and agriculture sector-such as animal or plant disease introduction or food contamination-could result in severe animal, plant, or public health and economic consequences.

The Pilot Study Group has agreed that a protection plan for food and agriculture infrastructure and resources must focus on planning and preparedness, as well as early awareness of an attack. Science-based surveillance measures were essential to recognizing a possible attack on the sector so that rapid response and recovery efforts could be implemented to mitigate the impact of an attack. A protection plan must also be coordinated closely with response and recovery plans.

3.5. Fifth Pilot Study Meeting: May 4–5, 2006, Helsinki, Finland

The main objectives of this meeting were to discuss the new method of Risk Assessment established in participating countries and activities of the new Food Safety. In this respect technical discussions were conducted concerning how risk assessment tools may be utilized in addressing problems related with food chain security in the participating countries. The method includes the linear mathematical model similar to that used by Google, the simplifying assumptions and the complex interdependencies and accumulating effects and risks handled. It enables assessing risk caused by interdependencies between technical infrastructures, basic services and threats. This model may also be suitable for food sector analysis.

These discussions included experts from NATO member countries as well as NATO partner countries including Russian Federation and Finland. Attendance and participation by representatives from NATO NATO-Food and Agriculture Planning Committee (FAPC) served to expand the perspectives of the group and to present examples of practical applications of risk assessment methodology for food chain security.

3.6. Sixth Pilot Study Meeting: January 26–27, 2007, Lisbon, Portugal

This meeting was focused on discussions and the development of recommendations and action items for (a) analysis of the results of the gathered data of the survey, and (b) to disseminate the scientific knowledge gathered and experience gained.

Changing trends in food safety and food chain security were discussed. It was agreed that the amount of foods consumed outside the home is increasing throughout the world and this changes the trends in food safety. Justification for the concerns with food chain security was presented; in particular the awareness that manuals for intentional contamination of food are widely available, that the use of biological or chemical weapons against the food supply could cause mass casualties and that even an ineffective attack could cause significant economic and psychological damage. Some documented intentional contamination events with food were referenced by the presenters. The vulnerability assessment methodology used by the FDA after September 11th was shared. In a first approach, most vulnerable foods and probable agents were identified and food/agent scenarios were designed resulting in initial risk rankings for types of foods. After this, CARVER (Criticality, Accessibility, Recuperability, Vulnerability, Effect, and Recognisability) plus Shock method was developed aiming at simplifying and standardising this process. It identifies foods for evaluation, assemble evaluation teams, develop flow diagrams for each product and develop terrorist profiles. Examples of terrorist profiles were given. FDA's Food Security guidance documents that represent the agency's thinking on appropriate measures that can be taken by difference segments of the food industry to minimise the risk of food being subjected to tampering or criminal or terrorist actions were presented. The basics of the ALERT initiative (Assure, Look, Employees, Report and Threat), which intends to increase the awareness of food defence among stakeholders was shared. The pilot study group agreed to join the efforts of FDA-USDA and EFSA in terms of food safety.

3.7. Seventh Pilot Study Meeting: October 25–26, 2007, Birmingham, United Kingdom

The meeting focused on recent real-life examples-case studies on food chain security. In this respect discussions included the recent frauds and effective countermeasures taken and the possible health hazards, the importance of communication on food terrorism and possible use of food as a vehicle for terrorism, the public confidence to government(s) and-or authorities. The case study was an example of how the surveillance systems, legislation and recall procedures function effectively to contain an incident or breach in food chain security by limiting the number of affected individuals.

The presentations on risk communication centered on the consumer perception of risk both with regard to deliberate (terrorist) threats and to accidental or inherent food hazards. A number of barriers have been identified which affect the way in which a consumer will respond to a particular message. Of particular importance for the work of the pilot study group was the matter of trust in the organization issuing the information. A number of components could be identified that impact or contribute to this trust and the pilot study group determined that significant cultural differences can be shown to exist in relation to them. Differences have also been identified in the way people use risk information, further complicating the issue of risk communication.

The role of the media and how it could be used by a terrorist organization was also discussed in detail. In this respect the issues surrounding the relationship between the media and government organizations trying to address (risk communication) was considered as a food safety issue. The presentations included consideration of some hazards which might be used in a deliberate contamination incident and what barriers and advantages these might have, depending on the objectives of the terrorist. It was concluded that the real consumer perception of food chain security hazards has to be taken into consideration.

The results of the gathered data were shared with the group and finalized. Despite the variations in data collection, sampling and language the preliminary results have revealed that (i) safety management systems are all implemented in the sample group, in some countries even 100%; (ii) crisis management and recall systems exist and are being used or tested and can respond very quickly.

3.8. The Eight Pilot Study Meeting Was Held on September 19–20, 2008 in Antalya, Turkey

This was the final meeting of this pilot study and it was solely devoted to the preparation of the final report and other outcomes of the pilot study. In this respect the pilot study group has agreed to divide the outputs into three main categories:

- A summary report to be submitted to the committee
- Country profiles in terms of food chain security
- Results and analysis of the data gathered via the questionnaire (nine countries 360 responses)
- NATO Series Book

4. Participating Countries

Pilot country: Turkey

NATO Members: Belgium (Pilot Study Co-Director), Bulgaria, Czech Republic, Germany, Hungary, Italy, Netherlands, Portugal, Romania, Spain, Turkey, United Kingdom, United States (Pilot Study Co-Director).

NATO Partners: Finland, former Yugoslav Republic of Macedonia,[1] Moldova, Russian Federation,

Mediterranean Dialogue Countries: Algeria

NATO/Science for Peace and Security web site for Food Chain Security Pilot Study (http://www.nato.int/science/pilot-studies/fcs/fcs-index.htm)

5. Conclusions

The Pilot Study was very successful and without NATO support it would have been impossible to have such an international network, co-operation and forum which allowed using a wide range of viewpoints and approaches mixing academic and public views. It developed excellent profiles of emerging issues in terms of Food Chain Security important to both NATO member and partner countries. The conclusions can be made as follows:

The best way to avoid or decrease the harms because of terrorism is to have an active and trustworthy relationship and co-operation between the public and private sectors. In order to exchange information between public and private sectors it is highly recommend creating a forum for Public Private Partnership (PPP) at every level.

The Pilot Study involved many stakeholders – government, military, research institutes, universities, and the private sector – in capturing the up to date technical information and the production of Internet materials giving the details of annual developments helped to disseminate the information in this rapidly developing field.

---

[1] Turkey recognises the Republic of Macedonia with its constitutional name

The Pilot Study served as a forum to foster and grow the network of Food Chain Security professionals from about 20 countries. This professional network proved to be an invaluable resource for information exchange and transfer, even during the periods between the meetings.

The study is important for end users because it will draw together various strands of the issue of food chain security and analyze them in an independent and scientific manner. Many current publications on the subject are not independent but published by persons or organizations with a vested interest (often financial) in maintaining heightened interest in the matter, whereas this pilot study and its outcomes do not have that handicap.

The pilot study created an international network of exchange of good practices and institutional framework in the field of food terrorism. The pilot study allowed confronting the threat of food terrorism to the current tools used in the traditional risk analysis.

During the pilot study, individual representatives of different aspects of the food chain were contacted in different countries involved, in order to answer the questionnaire. This phase was a kind of mobilization procedure to sensitize the stakeholder to the risk of food terrorism.

The group was very international, there were ample chances to learn from each other and then give the most valuable information to end-users. Most people are more familiar with food safety, not security; therefore the pilot study was a very good opportunity to fill this gap. The outcomes of this study would inform all the responsible authorities.

Main achievements include the dissemination of knowledge about the diverse situations of food chain defense throughout the countries of the participants. These pinpoint differences in attitude and preparedness. From these it may be possible to draw sets of minimal preparedness conditions required and also of different defensive schemes which might be adopted. Minimal requirements in detection will also be forthcoming, as will be recommendations for the shortest possible time for the various phases of detection and response. The latter elements constitute project deliverables.

When all people at all times have physical and economic access to sufficient, safe and nutritious food any matter that prevents access by the consumer to foodstuffs will be a security issue. This will range in severity from lack of access to foods of choice through to starvation from no access at all. Depending on the structure of food delivery (e.g. just in time) any interruption in that supply can quickly become an emergency, or appear as one.

This pilot study was dedicated to risk issues related to the food chain. In this perspective, Civil Emergency Planning (CEP) was at the center of the discussions and work done by the participants.

The possibility of an event affecting food quality/safety and especially the innocuousness of the supply constitutes a serious security issue, and might potentially affect whole or very high percentages of the populations in various countries. Risks increase with the degree of economical development. Contingency plan drawing is therefore a very important issue, and an issue which has to be tackled by the entities in charge of CEP, who will have to contact each other, as they have to, e.g. in case of radioactive events.

One promising tool to improve decision making process is Operational Risk Management (ORM). This approach prioritizes the sites that represent the greatest potential risks in terms of severity and probability, likelihood of an attack and the most effective use of financial/technical resources for remediation and conservation are allocated.

6. Recommendations

The issues related with Food Security will continue to be a priority topic for all countries as it is a global issue that is continually developing. The results of this pilot study are relevant for many partner countries that will be facing these issues. Many of the countries in the pilot study expressed a need to create additional fora and workshops to continue the dialogue on this issue among the professionals.

Acknowledgements

As the Reporter of the Pilot Study I would like to acknowledge all country representatives (authors) for supplying the relevant information to this chapter.

Food Chain Security and Vulnerability

Sébastien Brunet, Pierre Delvenne and Frédéric Claisse

Dépt. de Science Politique, University of Liege, 7, Bd du Rectorat, B 31, Bte 29, 4000 Liege, Belgium

Abstract

In our contemporary societies, the food chain could be defined as a macro-technical system, which depends on a wide variety of actors and risks analysis methods. In this contribution, risks related to the food chain are defined in terms of "modern risks" (Beck 1992). The whole national economic sector of food production/distribution is vulnerable to a local accident, which can affect the functioning of food chain, the export programs and even the political system. Such a complex socio-technical environment is undoubtedly vulnerable to intentional act such as terrorism.

1. Introduction

Food chain security is a very complex issue involving a wide range of stakeholders and experts. As feature of a pluralist society, such actors are not monolithic blocks (Renn 1998). According to Ulrich Beck (1992), risks related to the food chain are also a characteristic of our "risk societies". In other words, we (as consumers, citizens, policy makers or scientists) are confronted to risks which are: invisible to the "profane" (need for scientific expertise); spatially unlimited (need for international action and cooperation); temporally undetermined (need for a long-term action and foresight); scientifically under discussion (need for more scientific investigation to reduce uncertainty).

Consequently, the growing spatial and temporal interconnections of our modern societies increases the level of difficulty experienced in dealing with risk analysis and management (Adam 1996). The whole national economic sector of food production/distribution is vulnerable to a local accident, which can affect the functioning of food chain, the export programs and even the political system. Such a complex socio-technical environment is undoubtedly vulnerable to intentional act such as terrorism.

In this chapter, we will propose an analysis of the food chain security regarding the concepts of vulnerability to terrorism and modern risk.

H. Alpas and B. Cirakoglu (eds.), *Food Chain Security*,
DOI 10.1007/978-90-481-9558-9_2, © Springer Science + Business Media B.V. 2010

2. Food Terrorism: A "Modern Risk"

Risks related to food security are very diversified. Among them, we will focus on a specific type of risk that is food terrorism. For such risks, the intentional dimension is central and represents an archetypical situation in terms of risk analysis. We mean that the notion of modern risk, when applied to food chain security and terrorism, is somehow extended further than Beck's initial idea, which mainly points at the unconscious and unwillingness to produce those risks labeled as "modern" (Beck 1992). Hence, in case of terrorist acts the vulnerability of the food chain is the central key-point. In this section, we propose an analysis of terrorist risks related to the food chain in terms of "modern risks". As briefly introduced above, five main dimensions characterize those risks: time-space scale, uncertainty, invisibility, human origin and democratic feature of the damage. We suggest to focus on each of them in the following sections.

2.1. An Enlarged Time–Space Scale

Misappropriation and offensive use of biological pathogenic material against the food chain in order to reach objectives labelled as "terrorist" could have detrimental impacts on extremely large geographical areas likely to overpass state's territorial limits. Because of the globalisation process in which contemporary societies are involved, damages occurring at one specific point of the food chain can easily spread like wildfire. Especially, commercial partnerships and economic systems are speeding up the propagation of bad events. However, they are also likely to radically affect the fate of future generations by altering their environment in an irreversible way. Managing such kinds of risks therefore requires to implement preventive measures taking into consideration enlarged spatio-time scales. The international dimension of "modern risks" – among which we consider food terrorism – also challenges State's sovereignty. Moreover, it stresses the issue of the State's capacity to react promptly and adequately to such threats. Therefore, as in the case of other "modern risks" (nuclear, asbestos, climate changes...), the especially enlarged spatio-time scale characterizing food terrorism requires preventive and management measures to be implemented at a supranational level, for instance by regional organizations such as the European Union (E.U.).

2.2. A High Level of Scientific and Social Uncertainty

Like other "modern risks", food terrorism is characterized by a high level of scientific and social uncertainty. On the one hand, identifying food terrorist risks appears to be a very difficult task for which there are countless activities and products likely to be used and/or combined in order to reach destructive objectives. Scientific uncertainty regards here the issue of the elaboration and production of terrorist means applicable to the food chain. But there are also several uncertainties regarding both the modes of dissemination of such terrorist means in the environment and the limits of such dissemination.

On the other hand, such a scientific uncertainty reflects in a great social uncertainty. Indeed, such a "modern risk" as food terrorism obviously reduces the part of the universe which man thought so far to be totally under his control. We can link that up to Latour's consideration (2003) that the risk society is not a society where the risks are higher but a society where there is an increased awareness that a complete mastership over actions is a modernist fiction. But there are even worst consequences: daily and safe activities in Western societies which have been until now implemented by numerous people in there everyday life, suddenly appear as likely to be highly dangerous and out of control. For example, eating or drinking are now on the way to become activities that we suspect to provoke highly dangerous consequences.

2.3. *Invisibility of the Risks*

"Modern risks" – and in this case food terrorism – seem unlikely to be detected by the sole help of the five senses. For example, identification of contamination elements requires implementation of sophisticated scientific tools. Contrarily, without such scientific tools, there is a risk of both numerous false alarms and bad jokes. Invisibility of "modern risks" therefore results in making the role played by scientists unavoidable, although the latter is ambiguous: on the one hand, scientists are the only ones which hold the knowledge enabling to identify, evaluate and prevent from risk's occurrence; on the other hand, scientific community frequently seems to be at the origin of the risk itself because it permitted some "risky" technological progresses to develop. Such an ambiguous role crystallizes criticisms against scientists and industrials that are often accused to be at the origin of prejudicial situations, while (at least for the scientists' side) they are considered as the only ones able to manage and to solve such situations. In the case of the food security, risks are most of the time invisible for the laypeople and they need scientific mediation in order to be identified. In other respects, invisibility of risks to laymen's eyes stresses a situation of dependence and therefore underlines a balance of power between those who hold scientific knowledge and expertise and those who do not.

Therefore, such invisibility may provoke a strong social demand for "zero risk". In order to meet this social demand, a wide range of scientific tests can be implemented in order to reduce specific modes of contamination of the food chain. But this high level security has huge technical, social and economic costs for (most of the times) little impacts.

2.4. Human Origin of the Risk

Food terrorism as well as other "modern risks" obviously has a human origin since they result from specific scientific activities initiated within research laboratories or food industry. Instead of prejudicial natural phenomena like earthquakes, hurricanes, etc., the threat's origin actually rests on the human society itself.[1] It is interesting to pay attention to the fact that the human origin of "modern risks" is not intentional, while when it would be the case, the result is not always the one expected by the perpetrator of the risky activity. However, when it is to discuss food chain security as a modern risk because it represents a serious target for terrorists, the intentionality feature of a risky activity comes in. Food terrorism obviously has a human origin, and it can be treated as a modern risk because it relies on the very nature on the economic and technological contexts of food chain in a global world. This is then amplified by the very complex and interconnected organisation of the food system, which involves many actors both human (operators, veterinary services…) and not human[2] (test boxes, cooling systems…).

2.5. Democratic Feature of the Damages

Damages likely to be caused by a food terrorist attack do not affect specific and limited socio-economic or socio-professional groups. Instead of a socio-professional disease such as *silicosis* (for instance), which affected mainly – if not exclusively – mineworkers, food terrorism as well as nuclear accidents like a radioactive leak are likely to affect everyone notwithstanding socio-economic or socio-professional status. Therefore, such a risk can be avoided neither by implementing sophisticated preventive measures nor by leaving the social group to which one belongs and which is threatened by the risk. There is actually no possibility for an "exit" – in the sense of Hirschman words (Hirschman 1995). In the food case, this inescapable situation is reinforced by the fact that human beings cannot stop eating, and increasingly rely on food chain markets' organization and supply. This is an everyday life activity nobody can really avoid.

The concept of "modern risk" truly stresses the fact that traditional processes of risk identification, assessment and management are therefore out of date regarding several aspects directly linked to the characteristics we underlined here above.[3]

---

[1] Although numerous so-called "natural" disasters are increasingly linked to global warming, having itself a great probability of being due to human activity.

[2] We refer to the Actor Network Theory vocabulary developed by social scientists such as Michel Callon (1988) Bruno Latour (2005).

[3] According to the reflexive modernization theory (see e.g. Beck 1992; Beck et al. 1994; Pellizzoni 1999; Beck and Lau 2005; Delvenne and Rip 2009), the society goes along evolving processes but the modernist routines embedded in the institutions render the latter not able to deal with the threats' seeds contained in society's evolution (including terrorist threat).

The different degrees and aspects of uncertainty, either scientific or social,[4] now appear as factors to be taken into consideration by both private and public decision-makers who cannot just proceed as they did before to apprehend the risks.

On the basis of the above-mentioned characteristics, "modern risks" therefore stress the need for a new political organization requiring:

- Decision-making procedures taking into account enlarged spatio-temporal scales
- Demystification of scientific knowledge and expertise
- Opening up of expert and decision-making systems
- Reintegration within collective decision-making processes of some domains, which appeared so far as enjoying a huge autonomy beyond the democratic accountability and control – which means that they are not under the legal control of the different State's bodies. Along with military labs, one should also evoke here other research fields such as modern biotechnologies.[5]

3. Modern Risk and Vulnerability

Modern risks related to the food chain reveal both the vulnerability and the domination characterizing relations between our modern societies and their socio-political and socio-technical environment. On the one hand, with respect to the socio-technical environment, our societies are indeed highly dependent on several socio-technical networks and complex systems without which the functioning and the well being of our modern societies would be compromised. However, such dependence is closely linked to our modern societies' vulnerability to terrorist acts as it provides a lot of targets often quite easy to reach. Up to now, terrorist attacks through the food chain are not really privileged by terrorist groups maybe because traditional systems of risks analysis are sufficient or maybe because the diversification of the food chain does not allow huge impacts in terms of societal disruption. Or maybe just because food chain terrorism does not appear to them as the best target (compared to others like public places, transportation networks or capitalistic icons) to achieve their objectives.

On the other hand, the relation that our Western societies maintained with their socio-political environment is characterized by domination since their richness and prosperity are indeed grounded in an unequal balance of power between themselves and the other world's areas providing them with a huge part of their energetic, natural and other economic resources. However, such domination provides terrorism with the structural context in which it is often – if not always –

[4] From a constructivist point of view, scientific uncertainty is seen as the result of an ongoing process of negotiation, deliberation and decision among scientists. See B. Campbell (1995).

[5] Indeed, similar reasoning and programme could apply to modern biotechnologies, see S. Brunet (2001).

grounded. Assuming that it is the "failure of the imported State [...] that made the international regulatory mechanisms defunct while constituting an opportunity for substitute actors", Bertrand Badie is certainly right when he asserts that "the *hegemon* from yesterday had stabilizing virtues, but is losing them today, tending increasingly to generate instability" (Badie et al. 2005). Terrorist networks are contingent on social expectations, fed off frustrations and humiliations. Their answer to this situation consists in the creation of a form of uncertainty (a struggle based on the "wearing effect of time"), which threatens the fragile complexity, the organization, and the prosperity of Western countries.

Contemporary terrorist risk is therefore fully linked to some of our everyday life activities such as travelling (airplane, subway, train…), shopping or eating. Therefore, it is only by examining our own practices (both socio-political and socio-technical) – an opportunity that terrorism, amongst other topics, provides us with – that we may be able to better understand terrorism and, even more, to prevent it more efficiently.

However, there are successive stages in the development undertaken by the relation between our Western societies and their environment. For its highly emblematic feature of such a process that rests on one of the more acute examples of "modern risks", the case of the food terrorism will give us an opportunity to take a cross-cutting example: the complex socio-technical characteristics of the food chain.

4. Vulnerability and Socio-technical Environment

4.1. Networks and Vulnerability

If September 11, and the series of anthrax attacks that followed, made Western societies entering a new configuration, it relied partly on the fact that this put in evidence the extreme vulnerability of all the macro-technical systems and networks which now underlie the acceleration of money, people and goods exchange flows within a global economic system. The enemy can now strike anywhere, so long as he manages to filter silently and harmlessly into the architecture of a network, which because of its apparent good working order might lead the terrorists to reach their target. Postal services, public transportations, water supply networks and food chain – all of these macro-systems which today constitute the infrastructure of our daily life –, are indeed multiplying, through their complex and closely-knit nature, the critical points of entry that make them sensitive to aggression or hijacking.

In a certain way, terrorism also acts as if it uncovers a specific relation to our environment. It is indeed the main property of these macro-systems: their capacity to supply users with goods of all kind, to provide them with services by connecting them to a distribution exchange or just to other users. So the terrorist may take advantage of such a systemic property, and it is this very connectivity that makes his action possible. The modalities of such an action are numerous: it can consist in mingling with one of the units in motion, like e.g. bus, underground train, letter or postal parcel; in attacking the element of a network, like e.g. passenger station, airport, sorting office; in infiltrating the actual flow, like e.g. water supply service, distribution conveyor, food chain; or, even more directly, the central unit which organizes the flow. According to its objective and to the scale of its intervention, the terrorist act will aim at a wide effect (panic, demoralization, political pressure on the opponent, specific demand) on the system as a whole, either through a local incident or through a more global propagation, like for instance along flow channels of some agent put in charge of contaminating users or other units (e.g. computer viruses or their biological counterparts).

A food terrorist act could be successful in transforming an everyday gesture, eating or feeding, in a test of force, with all that is implied in terms of bodily commitment, perceptual wakefulness, suspicion and mistrust towards anything that could incite trouble (suspect smell, unknown producer or origin, unusual weight or texture) – all gestures which are perfectly reasonable but, because they had to be made in contexts usually not submitted to this kind of constraints and suspicions, arouse some fear that the population would plunge into collective hysteria, whereas the phenomenon may more probably and modestly be an unprecedented exacerbation of the latent potentialities which are enclosed in every socio-technical system and which are revealed in all of their dimensions by the terrorist act. The same process of testing could also concern other everyday gestures by which we connect by ourselves to a supply network, or filter as elements in motion of an interconnection system: turning the tap, going to the market, using public transportations.

Threats to the food chain are also symptomatic of the inherent weakness of complex and highly interconnected systems and networks. The different steps that food follows from the ground to the plate have never been so numerous and, for most of them, so obscure. Everyday, a lot of people ingest the major part of their nourishment without carefully wondering about its origin or its ingredients.[6] Despite the regulations and laws (such as mandatory traceability), the reality that lies behind the label remains hidden to the layperson. The open market creates opportunities for worldwide exchanges and, at the same time, for some legal and technical interstice. On the legal side, we are far from a global harmonization, while on the technical side each new intermediate may be a potential new security breach. One month and a half after September 11, some scientists "believed that the next major terrorist target [in the USA] would be the food supply".[7] In 2002,

[6] Notwithstanding the compulsory ingredient lists printed on the packages. It's also important to point out that a growing number of consumers seems to feel more concerned about a healthier and more ecological way of life, and are therefore more vigilant than the average towards food (among other things).

[7] "UNC Professor: Next Terrorist Target Could Be USA's Food Supply", CBC, October 24, 2001 http://www.wral.com/news/local/story/100208/

the World Health Organisation (WHO) identified food borne disease outbreaks and incidents, including those arising from natural, accidental and deliberate contamination of food, as major global public health threats in the twenty-first century.[8] Since then, though, there has not been any major case of food terrorism (at least officially and if you do not consider envelope containing Anthrax as food). However, WHO's scientists still consider it as a major threat, underlining the potentiality of great damages with very few contaminating material: for instance, one gram of botoxin (the organism that cause botulism) can kill hundreds of thousands of people. It also seems obvious that both the symbolic dimension of the act of contaminating food (a direct source of life) and the economic impact of a major food crisis play a role in the high ranking of food terrorism on the threats' scale and political attention. Hence, the food chain appears as a seductive target, allowing to reach a vast amount of persons, but at the same time remaining quite unaffected. And one could go further in assuming that giving it so much attention could stress its strategic importance and point it out as a target. This is particularly true if one considers that "the execution of an attack depends upon the attractiveness of the target and the terrorists' resources and plan" and that "to a terrorist, civilian populations; targets of historical, cultural, and national significance; and infrastructure that underpins the [US] way of life are all 'fair game'" (Garrick 2004).

We are thus forced to re-learn that our comfort, our power of action, the very routines of our daily life are to a great extent the expression of our vulnerability and dependence on incredibly wide and complex networks and socio-technical systems. What this exercise of reflexivity invites us to, is indeed a new experience of "collectiveness", an experience that French sociologist Nicolas Dodier chose to call technical solidarity.[9] One can remind too that the emergence of modernity is itself strongly linked to this increased dependence, which is in turn the corollary of an ever-growing capacity of control by man upon his environment, his peers as well as upon himself – in other words, everything that, according to Norbert Elias, is characteristic of the civilizing process. In such a way that our very existence as

---

[8] World Health Organisation, "Terrorist threats to food: guidance for establishing and strengthening prevention and response systems", 2008 edition. The first edition was published in 2002.

[9] "I suggest to call *technical solidarity* this form of link between beings created by the functioning of technical sets. The sociological strength of this solidarity, its ability to link human beings to each other beyond the point they can aim at through their actions, is all the greater since the objects have an autonomous functioning capability. The notion of network, with its implications on the relation to space, is even more relevant here. It supposes, broadly speaking, the existence of links, which are well enough consolidated between several technical objects, spread out through space. It was therefore necessary that the planet had first constituted a place filled with artefacts linked to each other according to long chains, before technical solidarity asserts itself as the medium of links of a new kind between people, crossing pre-existing groups, and partly liberated from the boundaries of ordinary space, due to new capacities of fast circulation of beings all along its paths" (Dodier 1995, pp. 1415). In another perspective, more strictly focussed on actual technical macro-systems and the new situations f dependence they induce, see Gras (1993).

individuals is also the expression of our interdependence on those systems and on the other individuals who are linked to them as we are.[10]

The awareness of this interdependence is an emerging phenomenon only recently observed by sociologists: the network is not only the form more and more often taken by the organization of our socio-economic activities but also, under the effect of a still deeper incorporation of information technologies, notably Internet, in the web of our daily life, the representation that accompanies this connexionnist model.[11] The network abandons the vocabulary of its architects and promoters, to become a cognitive point of reference, which is gaining more and more relevance to catch the nature of what could be called a new social configuration. Even better, it becomes a common form invested by partners of different kinds to coordinate their actions and to deal with problematic situations together.[12] If the date of September 11 represents something else than a handy journalistic landmark and truly functions, as we think it does, like a reconfiguration point, actually opposing a 'before' and an 'after', it is also due to the fact that the Western countries were, through these attacks, confronted to a relatively new sort of internal threat, in this sense that the logistics of the terrorists relies on the hidden aspect of our societies' power and wealth: the vulnerability of the socio-technical forms of their organization.

4.2. Irreconcilable Demands

From this point of view, it is obvious that the reduction of the terrorist risk to zero can only be imagined in the situation, definitely unimaginable, of the interruption or the suppression of a service and, consequently, of its distribution network. A biological threat of contamination of a water supply network, for example, or of a planned poisoning of a link in the food chain, can make necessary the improvement of the existing measures regarding sanitary watch and quality control, but as far as surveillance of the buildings, people and public places which could be potential targets of attacks is concerned, as well as the fight against the groups which are likely to organize such attacks, it is here a classical problem of political sciences that has to be raised again in new terms: the question of the capacity of the State to maintain and legitimate its monopoly of physical violence.[13] It is true that

---

[10] Interdependence (with no consideration however as to the technical component of our collectivities), is also a key-notion in Elias, who has been systematically rediscovered for a few years. See, of course, *The Civilizing Process. Sociogenetic and Psychogenetic Investigations*, Blackwell, 2000 (Reprint).

[11] See, in very different perspectives, Castells (2000) and Boltanski and Chiapello (1999).

[12] We are here employing the notion of form in a sense close to Thévenot's (1986).

[13] To put in perspective, as well as to give a new insight into what is at stake in the emergence of the "risk" category, when applied to classical questions of the sciences of politics, notably the functioning of State, see Lemieux and Barthe (1998).

September 11 blatantly revealed the weaknesses and difficulties experienced by public agencies, intelligence services and technologies of state control, in infiltrating worldwide terrorist networks and anticipating actions of inconceivably large scale, but also in communicating the information among themselves that probably needed years of preparation and coordination. Among the most dismaying failures, the communication interception network known as Echelon, was denounced, as it may have been (also) reassigned to missions of industrial espionage,[14] and was obviously neutralised or bypassed by the organizers of the terrorist attacks (not forgetting the possibility, even more overwhelming but unprovable, of collusions and internal complicities). Does this mean that the state as a form, in one of its most decisive prerogatives, is now outmatched by the flexibility and the plasticity of criminal networks or, more simply, that the complexity of our organizations and of our socio-technical systems furthers this kind of criminality more than at any other moment in history, by creating new vulnerabilities? And if the new network configuration makes us more vulnerable, how can we change this, unless, very conventionally, resorting to traditional forms of state control, intelligence gathering, public space monitoring and governing of the populations, all forms which in other respects turn out to be partially outdated by the internationalization and the process of decontainment of risks?

On this side too, considering the way that the security of some systems was increased in the months that followed the September 11 attacks, and normative texts adapted in a great rush to make the country feel safer and virtually able to face an enemy now likely to come out from within, we must admit that we do not find out any sign of evolution towards a genuine reflexive work on our society's operation mode, just as nothing came out yet in terms of awareness of some of the effects of our domination over the rest of the world – save for a politically dominated part of the American intellectual circles (see the "Letter from American citizens to their friends in Europe" and the "not in our name" movement[15]). At first sight, the Western societies under threat of terrorist risk have not integrated yet the reflexive dimension carried by these risks, which put to the test pre-established arrangements and systems of all kind. Seen from the point of view of our relation to our socio-technical environment, the awareness of this vulnerability only acts for the moment as a reinforcement of previous cognitive patterns, resulting only in attempts at conciliating contradictory demands: maximal connectivity and mobility as to the regulation of exchange flows, but maximal security and control in the government of people and areas.

The capacity of networks to reconfigure themselves is nevertheless far from representing a drawback or a simple additional factor of risk. The mobility of units

---

[14] See the final study published by the Scientific and Technological Options Assessment Programme (STOA) for the European Parliament, entitled "Interception Capabilities 2000" (volume 2 of a larger set of publications about the "Development of surveillance technology and risk of abuse of economic information"), a rare, well-informed and reliable description of the potential risks related to Echelon, prepared by Duncan Campbell.

[15] www.notinourname.net

along a flow channel can even be turned into an advantage and constitute the very condition of control of these units. The prevention and management of all sorts of risks, be they industrial, sanitary, or related to the food chain, have succeeded, with the recent figure of tracking, in developing a clear example of compromise between both requirements of movement and security[16] – and tracking can itself be described as the equivalent form, for advanced modernity and network society, of what surveillance and discipline were for industrial societies in the previous configuration. Ensuring the identification, the follow-up and the localization of any unit in motion in a flow can represent an appealing alternative to a model of quarantine or *cordon sanitaire* which is doomed to failure when confronted to the intensification and the diversification of trade in a logic of network (as illustrated by the European management of foot-and-mouth epizootic disease in spring 2001).

5. Conclusion

Propagation of terrorism to technical macro-systems such as the food chain is key illustration to understand our contemporary relation to vulnerability and domination. Confronted to their own vulnerability, it is nothing but understandable that the dominants have the temptation to respond to a terrorist threat by the very means of their domination, without really realizing that by doing so, far from mitigating a risk, they multiply the chances of harmful events to occur. However, like any modern risk, food chain terrorism at least provides those subjected to it with the opportunity of a reflexive return to the elements, which lays at the origin of the threat, within both the context framing their own practices and the relation they keep up with their environment in the general sense of the word. If they do not grasp that opportunity, the odds are that the Western countries will continue to "pay the price of their hegemony" – unless, cynically, this price is considered to be low enough to, if not question the structure of the balance of power, at least take note of its effects. Without subscribing to the generous utopian view of a reflexivity extended to the international arenas where the general frames of the fight against terrorism are negotiated and defined (a view, however, which possibly lays at the heart of this paradigm in its performative will of societal change), we believe that it is only if we take truly the terrorist risk seriously in all the social, technological and political dimensions that feed its existence and its expansion, that the forms of an optimum treatment will take shape, on a long-term basis, of a paradoxical phenomenon which is deeply rooted in and revealing our mode of presence to the world. Food terrorism gives us the opportunity to (re)-think the global distribution of food and its socio-economic roots.

---

[16] About tracking as a new technique of government, and the specific difficulties related to its setting up, applied to three cases (Mad Cow disease, AIDS contaminated transfusion blood, GMO), see Torny (1998).

References

Adam, B. (1996). Re-vision: the centrality of time for an ecological social science perspective. In S. Lash, B. Szerszynski and B. Wynne (eds) *Risk, Environment and Modernity*. London: Sage, pp. 84–103.

Badie, B. et al. (2005). *Dans Quel Système International Vivons-Nous?, Qui a Peur du XXIe Siècle? - Le Nouveau Système International*. Paris: La Découverte, pp. 14–15.

Beck, U. (1992). *Risk Society: Towards a New Modernity*. London: Sage.

Beck, U., Giddens, A., and Lash, S. (1994). *Reflexive Modernization*. Cambridge: Polity Press.

Beck, U. and Lau, C. (2005). Second Modernity as a Research Agenda: Theoretical and Empirical Explorations in the 'Meta-Change' of Modern Society. *British Journal of Sociology* 99(4), pp.525–557.

Boltanski, L. and Chiapello, E. (1999). *Le Nouvel Esprit du Capitalisme*. Paris: Gallimard.

Brunet, S. (2001). Risques réflexifs et processus délibératifs. In Hupet P. (ed), *Risques et Systèmes Complexes. Les Enjeux de la Communication*. Bruxelles: Peter Lang, pp. 39–56.

Brunet, S., Claisse, F., and Rogister Y. (2005). From modern risks to reflexive society. An alternative approach to (bio)terrorism. In Michel Q. (ed.), *Terrorism: Cross Analysis*, Bruxelles: Peter Lang, pp. 71–104.

Callon, M. (1986). Some elements of a sociology of translation: domestication of the scallops and the fishermen of St Brieuc Bay. In John Law (ed.), *Power, Action and Belief: A New Sociology of Knowledge*. London: Routledge & Kegan Paul.

Campbell, B. (1985). Uncertainty as a symbolic resource in disputes between experts. *Social Studies of Science*. Vol.15, pp. 429–453.

Delvenne, P. and Rip, A. (2009). Reflexive Modernization in Action: Pathways of Science and Technology Institutions. *Theory, Culture and Society*, submitted for publication.

Castells, M. (2000). *The Rise of the Network Society*. Oxford: Blackwell (2nd ed.).

Dodier, N. (1995). *Les Hommes et les Machines. La Conscience Collective Dans les Sociétés Technicisées*. Paris: Métailié.

Elias, N. (2000). *The Civilizing Process. Sociogenetic and Psychogenetic Investigations*. Oxford: Blackwell (Reprint).

Garrick, B. J. et al (2004). Confronting the Risks of Terrorism: Making the Right Decisions. *Reliability Engineering and System Safety*, 86.

Gras, A. (1993), *Grandeur et Dépendance*. Paris: PUF.

Hirschman, A. (1995). *Défection et prise de parole*. Paris: Fayard, coll. "L'espace du politique" (French translation of *Exit, Voice and Loyalty*. Cambridge, MA: Harvard University Press, 1970).

Latour, B. (2003) Is Re-Modernization Occurring–And If So, How to Prove it? A Commentary on Ulrich Beck. *Theory, Culture and Society* 20 (4).

Latour, B. (2005). *Reassembling the Social: An Introduction to Actor-Network-Theory*. Oxford: Oxford University Press.

Lemieux, C. and Barthe, Y. (1998). Les Risques Collectifs Sous le Regard des Sciences du Politique. Nouveaux Chantiers, Vieilles Questions. *Politix*, 44, pp. 7–28.

Pellizzoni, L. (1999). Reflexive Modernization and Beyond: Knowledge and Value in the Politics of Environment and Technology. *Theory, Culture and Society* 16(4), pp. 99–125.

Renn, O. (1998). The Role of Risk Perception for Risk Management. *Reliability Engineering and System Safety*, 59, pp. 49–62.

Thévenot, L. (dir.) (1986). *Conventions Economiques*. Paris: PUF, Cahiers du CEE.

Torny, D. (1998). La traçabilité comme technique de gouvernement des hommes et des choses, *Politix*, 44, p. 51–75.

The Contribution of Food Safety Management Systems to the Prevention of Deliberate Food Contamination

Madeleine Smith

Chemical Engineering, University of Birminham, Edgbaston, Birmingham, UK, B15 2TT

Abstract

Food Safety Management systems such as HACCP can help to prevent deliberate contamination of food. These systems are widely used already by the food industry to prevent accidental contamination. This makes them an attractive and cost effective way to address the issue of deliberate contamination.

1. Introduction

A reliable and continuous supply of safe food is essential for food security. Vulnerabilities in the food chain must be identified, as far as possible quantified and then risk rated to ensure adequate controls are implemented and resources targeted. The deliberate contamination of food to render it unsafe for human consumption is an obvious area of concern and it is prudent to consider this in some detail. There are a number of other matters that may also threaten food chain security. Examples include:

A disruption of the country's infrastructure which prevents food from reaching consumers (DEFRA 2006)

Problems of animal or plant health which destroy food sources (DEFRA 2009; Crutchley et al. 2007)

Incidents which make food unsafe through radioactive contamination such as occurred in Welsh sheep after the accident at the Chernobyl Nuclear power plant (Blake and Harvey 2007)

These important matters are considered in detail elsewhere e.g. Orre (2005), DEFRA (2006, 2009) and will not be discussed further in this chapter. Instead the focus will be on the potential for deliberate contamination of foodstuffs.

H. Alpas and B. Cirakoglu (eds.), *Food Chain Security*,
DOI 10.1007/978-90-481-9558-9_3, © Springer Science + Business Media B.V. 2010

Unsafe food is defined by article 14 of Regulation (EC) No 178/200 as food which is injurious to health or unfit for human consumption. Food which is injurious to health is self explanatory – it is food which causes some damage to the consumer, such as a food borne illness, an intoxication or physical damage such as cuts, abrasions or broken teeth. The same legislation explains that food can be considered unfit if it is

'Unacceptable for human consumption according to its intended use, for reasons of contamination, whether by extraneous matter or otherwise, or through putrefaction, deterioration or decay.'

Food which is contaminated with something that is unacceptable, for example a non hazardous chemical which taints the product, would render the food unfit rather than injurious to health but still, by definition, unsafe. For the purposes of this chapter, the definition of unsafe food will encompass contamination from either category.

2. Contamination

Food may become contaminated at many points during the food chain. Contamination can be accidental or deliberate. A contamination incident can be very significant with substantial morbidity and mortality. Examples include the outbreaks of *E. coli* O157 in Sakai, Japan where over 8,000 school children were affected (Michino et al. 1999) and Toxic Oil Syndrome in Spain (Sanchez-Porro Valades et al. 2003) which affected 20,000 individuals, at least 2,500 of whom died. The majority of food contamination incidents (including the above mentioned) are inadvertent and the frequency and unpredictability of their occurrence has led some researchers and officials to conclude that similar effects could be caused by deliberate contamination (Bénoliel 2007; Khan et al. 2001; Crutchley et al. 2007). Careful consideration of examples from both accidental and deliberate incidents is valuable. By so doing, the actual risk of deliberate contamination can be assessed and appropriate controls identified.

2.1. Characteristics of a Contaminant

In order to cause a breach in food chain security, i.e. to make the food unsafe, a contaminant must have certain characteristics. By studying accidental contamination incidents one can draw up a list of relevant criteria that a contaminant needs to cause such a security breach. These will include the following points:

The contaminate must be hazardous to humans.

The contaminant should not be noticeable – i.e. it must be tasteless, colorless, odorless and inert with regard to the food chemistry and composition. A contaminant that alters texture, taste, smell or appearance will be unlikely to cause a major incident because it will be noticed at the point of ingestion if not before (e.g. Fowle et al. 1996).

The contaminant should be stable in the food over the shelf life.

A contaminant having the intrinsic characteristics mentioned above has the potential to cause a food security breach. In order for a breach to actually occur the contaminant must also be introduced into the food. The significance of the incident then depends on the extent of the contamination. Maximum effect will be obtained if the following parameters also apply:

There is uniform mixing of the contaminant throughout the food. This will ensure it affects all consumers. Non homogeneous but widespread contamination can also have a similar significant effect.

The contamination occurs in food produced in large batches.

The food is a common, popular and/or inexpensive food.

The food has a short shelf life resulting in rapid consumption.

It is common to divide food safety hazards into three groups, biological, chemical and physical, and to consider each group separately.

2.2. Biological Contaminants

There are many published descriptions of accidental microbiological contamination of food. There is no doubt that these outbreaks, which are far too common, have caused significant illness and sometimes death, for example the outbreak of *E. coli* 0157 in Lanarkshire in 1996 (Cowden et al. 2001). The public health impact of these incidents has led to concern that bacterial pathogens or their toxins could be used effectively to cause deliberate contamination events (e.g. Crutchley et al. 2007) but in fact there are few published examples of such deliberate attempts. One reason for this may be the difficulty in acquiring sufficient of a suitable pathogen.

2.2.1. Bacterial Contamination

Two reports published in 1997 involve the deliberate addition of pathogenic bacteria to ready to eat food and it is important to critically evaluate these examples. Kolavic et al. (1997) describe how an outbreak of *Shigella dysenteriae* type 2 at a medical centre in Texas was investigated and determined to be both food mediated and deliberate. This conclusion was based on the epidemiological and typing evidence. The implicated foods were muffins and doughnuts, unusual vehicles for the transmission of food pathogens under normal circumstances.

However, the very low infectious dose of *Shigella dysenteriae* and the speedy consumption of the products apparently resulted in a 100% attack rate. A second report, also published in that year but describing an incident which occurred over a decade earlier (in 1984), describes how deliberate microbiological contamination affected the public when members of the Bhagwan Shree Rajneesh cult contaminated either the salad bars or the coffee creamers in at least ten restaurants in The Dalles, Oregon. The contaminant was an isolate of *Salmonella* Typhimurium, and 751 affected people were identified by the outbreak investigation (Török et al. 1997). The authors of the Oregon outbreak report suggest that the method used by the Bhagwan Shree Rajneesh cult could be easily replicated and that copy cat attacks are possible, but admit that none have actually been identified.

2.2.2. Parasite Contamination

Phills et al. (1972) report a singular incident in which some students at McGill University in Montreal, Canada were 'maliciously' fed parasite ova and this should also be considered an example of a deliberate contamination incident using biological material.

2.3. Chemical Contamination

2.3.1. Direct Contamination

There are documented cases of chemical contamination of food. A very well known case is that of the Toxic Oil Syndrome reported in Spain in 1981 (Sanchez-Porro Valades et al. 2003). This contamination incident affected more than 20,000 individuals and, by 2001, was considered responsible for over 2,500 deaths. The rapeseed oil which caused this incident was toxic because it was originally intended for industrial uses and as such had 2% aniline added to it. This reacted unexpectedly in the oil causing it to become toxic. Had it been used industrially this presumably would not have mattered but unfortunately it was sold for human consumption in Central and North-western areas of Spain. While this is not a deliberate contamination event, so the steps of acquiring the contaminant and introducing it into the food are not relevant, it is an example where contaminated food has been deliberately delivered for human consumption. The motive was presumably economic. The vendors sold the product door to door in unlabelled 5 l containers (Tabuenca 1981). Unregulated sale of foods, especially at competitive prices, has the potential to affect as many consumers as the vendors can reach and could be effective in a deliberate incident.

2.3.2. Indirect Contamination

Another well known example of chemical contamination is that of the contamination of Belgian animal products with PCB's and dioxins. This was an indirect contamination incident in that animal feed was affected. Contaminated mineral oil had been used in its production. Subsequently the farm animals ate the feed, the PCB's and dioxins were absorbed by the animals and so into the human food chain. This incident was considered to be accidental but has significant points. Only a very small amount of contaminant was used (approximately 1 g of dioxin into 50 kg of contaminated oil) (Van Larebeke et al. 2002) but because its dispersion could not be tracked or identified, all Belgian products of animal origin that could have been affected, including meat, poultry meat, meat products, meat preparations, milk, egg and egg products, were subjected to an EC Decision and prohibited for human consumption (EC Decision 1999/449/EC). This contamination incident had no impact on human health and the prohibition was lifted within a short time (just over a month for milk products (JFSS 1999)) but it caused substantial disruption for the food industry. There was, of course, also an economic impact. According to Buzby and Chandran (2003), the Belgian Government estimated the overall cost of the crisis to be 456 million euros.

2.3.3. Deliberate Contamination

A further example of how extensively a chemical can contaminate the food chain is the incident(s) of sudan dye contamination that occurred in 2003–2005 in the UK. Sudan Dyes (Sudan I, II, III & IV) are red colorants (azo-dyes) that are prohibited, under the Colours in Food Regulations 1995 and Directive 94/36/EC, for use in foodstuffs to be placed on the market in the EU. The dyes cause liver and bladder tumors in laboratory animals (Anonymous 1975) and in vitro studies on Sudan I suggest its carcinogenic potential applies to humans as well (Stiborova et al. 2002). Testing by French authorities identified contamination of imported hot chilli powder with Sudan I in May 2003. This caused certain emergency measures to be implemented and member states were asked to sample potentially contaminated products to determine the extent of the problem. It was found that dried chilli, chilli powder and some curry mixes were affected. Some products in the UK tested positive and were taken off sale. The Food Standards Agency issued guidance to businesses asking them to recall and withdraw any products containing the contaminated spice. Because chilli powder is an ingredient in many sauces, spice mixes, processed and ready to eat foods, this involved a substantial number of products. The investigation continued through 2003, 2004 and into 2005. By March 2005, 580 food products had been identified as containing the illegal dye and had been recalled from sale. This was the largest recall of foodstuffs in the UK (Anonymous 2007). The contaminated ingredient met many of the criteria for a successful contaminant – it did not alter the organoleptic properties of the product in a negative way (it enhanced the colour), was stable, widely used in long life products (some of which were then used as ingredients in other processed foods) and mixed well throughout the food stuff. The popular press suggested the recall associated with this contamination cost £100 million, although there is no confirmation of this figure available (BBC 2007).

In the sudan dye incident described above, an illegal contaminant was deliberately added to a food stuff. The motives have not been reported but as the dye enhanced the natural properties of the chilli (made it a brighter red) one may assume the addition was made for commercial reasons.

Another such contamination incident with more serious consequences was the contamination of Chinese milk products with melamine. Some food products may have a specification associated with their placing on the market, for example a minimum protein content. Protein levels in milk vary according to a number of factors (see, for example http://www.ruralni.gov.uk/index/publications/information_booklets/high_protein_system/benefits.htm) and a low level may need to be boosted by the addition of skim milk powder or other appropriate additives. The protein content of a food is determined by analysing its nitrogen content, so any nitrogen containing additive can appear to boost the protein content of the food by increasing the amount of nitrogen available for analysis. Techniques cannot distinguish whether the nitrogen is from protein or non-protein components. Melamine is an organic compound which may be used in the manufacture of plastics (Tyan et al. 2009), including food packaging materials (Lu et al. 2009). The compound contains 66% nitrogen ($C_3H_6N_6$) (Wu et al. 2009). Melamine is certainly not a legal food additive as it is toxic, affecting the kidneys and bladder in mammals unfortunate enough to ingest it (Gossner et al. 2009; Tyan et al. 2009; Dobson et al. 2008). Morbidity rates can be high, according to the level of contamination, and fatalities are usually due to acute renal failure (Gossner et al. 2009; Dobson et al. 2008). It is therefore a completely unacceptable component in any foodstuff. In the spring and summer of 2008, it became apparent that a number of infants in China were presenting with serious renal symptoms. The investigation identified an association with infant formula produced by a well known Chinese manufacturer (Wu et al. 2009). It appears that the milk used as an ingredient in this formula had been adulterated with substantial amounts of melamine, added to boost the apparent protein content (Chen 2009). Ultimately 294,000 cases were reported and six children died (Gossner et al. 2009). The contaminated milk was also used to produce other foodstuffs. Some of these were exported. Forty seven countries reported finding melamine contamination in products which had been imported from China, imported via third countries or illegally placed on their markets (Gossner et al. 2009).

The example of contaminated oil described in Section 2.3.2 was considered to be accidental. The sudan dye and melamine incidents were deliberate but done for commercial gain, possibly in ignorance of the consequences. There is one incident of chemical contamination which has been reported and is considered to be a deliberate incident in the same manner as the *Shigella* and *Salmonella* incidents referred to in Section 2.2.1, i.e. designed to deliberately harm consumers. This was reported in MMWR in 2003 and describes the intentional contamination of 200 lb (91 kg) of minced beef with a nicotine containing insecticide (Boulton et al. 2003). In common with the case reported by Török et al. (1997) the contamination occurred in a food premises selling to the final consumer. This is in contrast to the other two cases of deliberate microbial contamination (Kolavic et al. 1997; Phills et al. 1972) where the contamination occurred in private circumstances where one assumes the consumers were known to the perpetrator.

Boulton et al. (2003) report that the contaminated meat was sold over a 2 day period and consumers began to report illness after 24 h. Symptoms included:

Burning sensation to lips, mouth or throat, dizziness
Nnausea, vomiting, abdominal pain, diarrhoea
Sweating, blurred vision, headache, body numbness
Unusual fatigue or anxiety, insomnia
Tachypnea or dyspnea, and tachycardia or tachyarrythmias

The outbreak control team identified 92 people who had been affected and a further 16 possible cases. On investigation a company employee was arrested and indicted.

Another occasion where food is alleged to have been deliberately contaminated using chemicals involved a major supermarket chain in the UK. In July 2007 an unemployed man with a gambling addiction tried to extort money from the company (Williams 2008) by threatening to contaminate yogurt with caustic soda (BBC 2008). Caustic soda can be acquired for use in unblocking drains and can be purchased in small quantities without special control. It would certainly be a dangerous contaminant. However, there is no evidence that the blackmailer actually carried out the contamination in this example – he just claimed to have done so.

A further, rather strange, contamination incident was reported the following year in the Gloucestershire area of the UK. In this case food and wine on sale in two supermarkets were sprayed with a faeces and urine mixture (BBC 2009). The perpetrator carried a spray bottle containing the effluent into the store and used it to contaminate the food. One interesting point about this urine spraying case is that the perpetrator, Mr Daifallah, was allegedly a chemist by training (This is Gloucestershire 2009) but still chose to use body fluids for his contamination event rather than trying to purify some less noticeable but more toxic contaminant.

2.4. Physical Contamination

Food may also contain physical contaminants. These can cause damage to the consumer in a number of ways, according to the material. Typically these contaminants are either part of the raw ingredients e.g. stones, pips or stems, bones etc or have originated in the processing environment. Parts of the equipment, the premises, the operators or the packaging may all find their way accidentally into foodstuffs. These are very common incidents in the food industry. Table 1 gives examples of 20 alerts that were issued by the Food Standards Agency for physical contamination events in the UK during the period Nov 2007–April 2009. These were all considered to be accidental. The contaminants are all common products–glass, metal, plastic etc. In theory these could be acquired very easily by persons wishing to contaminate the food chain. Historically there have been incidents where malicious contamination has been reported to have occurred. One occurrence dates from1989 when baby food made in the UK by H. J. Heinz was allegedly contaminated with glass. The company apparently received a blackmail demand related to the incident (Lohr 1989). Other incidents of deliberate food contamination also occurred at this time, for example, an incident in Birmingham, UK which involved a major take away chain. A disgruntled employee, left to close down the premises after trade had ceased decided to add his own faecal material to a container of root beer ready for dispensing to the public (J. Millward 2010). This could be categorised as physical and/or microbiological contamination but the relevant point is that commonly available material was used as the contaminant. Such incidents are rare but have potential for disruption of the food chain and should be included in the determination of appropriate controls.

3. Contamination of the Food

3.1. Access to the Contaminant

The acquisition of a suitable contaminant is the first step for any person wishing to initiate a deliberate breach of food security.

3.1.1. Bacterial Pathogens

What is important about both the deliberate incidents described above where a microbial pathogen was used is that they both involved competent microbiologists working in equipped laboratories with access to pathogenic strains. The Bhagwan Shree Rajneesh commune housed a clinical laboratory and the medical centre implicated in the Texas outbreak stored frozen stock cultures for use in research and other activities.

Although pathogenic organisms can, in theory, be isolated from raw meat or soil, such an ad hoc approach does not seem so far to have met with great success as a source. Authors such as Bellamy and Freedman (2001) citing Henderson (1999), Bossi et al. (2004) and Arnon et al. (2001) allege that the Aum Shrinrikyo cult attempted to isolate pathogens from soil (anthrax and *C. botulinum*). They were unsuccessful (according to Arnon et al. 2001 because their microbiological technique was faulty) and instead perpetrated their terrorism in Tokyo using sarin gas (Okumura et al. 1996).

3.1.2. Microbial Toxins

Various studies have considered the risk associated with botulinum toxin based on the idea that access to the toxin itself can be achieved without having to culture the organism. The existence or development of botulinum toxins in ready to eat food is certainly a very serious food safety hazard (O'Mahoney et al. 1990).With the popularisation of botox injections for cosmetic purposes and the development of other medicinal uses for the toxin (Gessler 2005), laboratories are now producing it commercially. However Arnon et al. (2001) suggest that therapeutic botulinum toxin is unlikely to be a realistic source for deliberate contamination events as the licensed preparations contain such small amounts (0.005% of the estimated lethal oral dose per vial). Further, Bigalke and Rummel (2005) report that the more pure the toxin is, the less toxic it appears to be when ingested. The toxin molecules created by the microbe *C. botulinum* are associated with other proteins (Gessler 2005) forming a complex which protects the toxin from degradation by stomach acid and proteases (Bigalke and Rummel 2005). Highly purified toxin for therapeutic purposes could well be less attractive as a source for deliberate food contamination for this reason as well. To produce the poisonous neurotoxin complex for use in a contamination event, the perpetrator would, according to Bigalke and Rummel (2005), need 'a well equipped microbiological laboratory and skilled techniques'. Such facilities may be difficult for individuals or proscribed groups to access but will be available in states researching the use of biological weapons. According to Arnon et al. (2001) Iraq was responsible for producing substantial quantities (19,000 l) of concentrated botulinum toxin which, the authors state are 'not fully accounted for'. Zilinskas (1997) also considers the production of this botulinum toxin by Iraq. This material should have been destroyed according the UN Security Council Resolution 687 but Zilinskas explains that UNSCOM are unable to confirm that this destruction did actually happen. The implication is that some, or all, of the toxin could have been transferred to groups with malignant intent. If this is indeed the case, the toxin will still have to be transported from wherever it is being stored to the point of contamination.

3.1.3. Pathogenic Viruses

Viral contaminants associated with food borne illness are even more difficult to culture than bacteria, requiring cell culture methods (Straub et al. 2007) rather than liquid or solid media. Some have only relatively mild symptoms and/or other routes of infection (e.g. norovirus) making them unreliable as deliberate food chain contaminants, in spite of their efficacy in accidental incidents.

3.1.4. Parasites

The incident described by Phills et al. (1972) involved a biological agent as a contaminant. Unfortunately the authors mainly report the medical aspects of the cases (which were extremely unpleasant for the four people affected) and do not provide any information on how the incident occurred or whether a researcher was involved. However, it is difficult to see how else the perpetrator could have acquired the parasites, other than from a research or teaching collection. There was, of course, a much lower awareness of health and safety issues and of security in general at the time of the incident (1970) so it is possible that a student or other member of the university could have been responsible. Access to laboratory specimens and supplies is more tightly controlled now than it would have been in the late 1960s and early 1970s, making the acquisition, theft or removal of hazardous material much more difficult.

From an examination of the literature it would seem that unlike the bombs used to attack the London Underground in July 2005 (MP press release July 25, 2005), food related pathogens such as *Salmonella*, *E. coli* 0157 and *C. botulinum* cannot easily be produced in a domestic setting. Specialist facilities and expertise are required. These are expensive and time consuming to set up, require a certain infrastructure for support and, depending on their location, may be subject to registration and inspection. However, some uncontrolled sources of food borne pathogens or their toxins may exist, as indicated above. It would appear, therefore, that while being able to acquire sufficient pathogens or toxins to deliberately contaminate food is actually quite difficult to achieve, it cannot be completely ruled out.

3.2. Chemicals

The insecticide causing the minced beef contamination incident described in Section 2.3.3 (Black leaf 40) was not used on the premises by the business. It was presumably carried in on purpose by the perpetrator. Insecticides using nicotine are carefully controlled. According to an editorial note in the report this product had been de-registered in the USA for 10 years prior to the incident. There are stringent controls on pesticides within the European Community as well (Council Directive 91/414 EEC). Nicotine is also now banned as a component in pesticides in all member states. Commission Decision 2009/9/EC withdrew authorization for its use and required that existing stocks should be used within one growing season. However leftover or unregulated sources could clearly still be accessed in the USA and the possibility may exist in the European Union as well.

Nicotine can also be extracted from other sources, namely tobacco and cigarettes. Recipes for dilute solutions can be easily found on gardening websites (e.g. http://lists.ibiblio.org/pipermail/pbs/2003-February/012651.html) although it is illegal within the European Union to produce pesticides in an unregulated manner. The toxicity of the product made from these recipes would be hard to gauge but there is some evidence that attempts have been made to home produce such toxins for nefarious reasons. In London in April 2005 Kamel Bourgass was found guilty of conspiring to commit a public nuisance by use of poisons (CPS Press Release 2005). There was much discussion in the popular press about this case, namely the exaggerated use politicians made of it for their own ends, sometimes based on incorrect preliminary information. The media and politicians focused on the Mr. Bourgass's ability to make ricin and those criticizing the case pointed out that the evidence for this was limited to the existence of recipes, common kitchen equipment and a memo from the Algerian security police who had interviewed an acquaintance of Mr. Bourgass. According to Carrell and Whitacker (2005) 'there was no ricin'. Whilst agreeing with the authors that politicians should consider carefully what evidence they use to frighten the public and push through their own agendas, the case against Mr. Bourgass is important for this discussion in that what was found in his London flat was a jar containing traces of nicotine together with recipes for its extraction. According to the investigating officer, this is beyond dispute and was not challenged by the defence. Clearly Mr. Bourgass was aware of the toxic nature of the substance and had attempted to acquire some, presumably by extracting it from a nicotine containing product such as cigarettes. There was not very much in the jar and the issue for consideration here is whether home extraction is a plausible method for acquiring toxin to use in a food chain incident. While extracting enough nicotine to kill greenfly as suggested by gardening websites is probably quite easy, extracting enough nicotine for a major contamination incident would be far more difficult. Solarino et al. (2010) report the estimated lethal dose of nicotine to be 0.5–1 mg/kg of body weight. Each portion of food would therefore need to be contaminated with 30–60 mg for maximum effect. The dose in a cigarette is estimated to be between 1 and 1.2 mg according to the brand (DOH 2002), although Solarino et al. (2010) and Hagiya et al. (2010) both report higher levels. Extraction in any circumstances is unlikely to provide 100% recovery, but especially in uncontrolled circumstances such as a domestic kitchen. From these figures, then, it would seem that while it might be feasible to extract enough toxin to poison an individual or even the equivalent of a family unit by amateur means, boiling up cigarettes is probably an ineffective way to acquire sufficient toxin for a major food chain contamination incident.

3.3. Physical

Considering the examples above, it would seem that physical contaminants such as glass, plastic etc may be readily available to someone wishing to deliberately contaminate the food chain. By comparison acquiring enough of a suitable biological or chemical contaminant to cause a major food chain incident is not easy to achieve by unskilled workers in a domestic setting. However some potential (unregulated) sources can be identified, e.g. illegal pesticides or other products that have escaped destruction, inspection or control, so with this in mind and given that documented (albeit rare) examples of the deliberate contamination of food exist, it is worth considering the next stage in a potential contamination incident which is introducing the agent (physical, chemical or biological) into the food.

3.4. Access to the Food

Any person wishing to deliberately contaminate food must be able to access the product in order to introduce the contaminant. In some of the deliberate contamination incidents examined above the contaminant was introduced into the food just prior to consumption. In two cases this was at private premises (*Shigella* on cakes and *Ascaris* ova in celebratory meal). In the other examples retail or catering premises were chosen. The advantages in using this point of access for the perpetrator are that:

The food will not be subjected to further processing (e.g. heating) which could inactivate the contaminant.

Dilution of the contaminant is minimised.

Also, there must be public access to the food in retail premises and, to a more limited degree, in catering premises. Public access isn't essential if the perpetrator is an employee as in the Black Leaf 40 or root beer incidents. The disadvantage of contamination at retail (for the perpetrator) is that generally only small amounts of food can be accessed and mixing the contaminant through the food may be problematic. Access to food earlier in the food chain could result in a more widespread contamination incident.

In the UK as in many other countries, food is provided to the final consumer via a supply chain that tends to involve a number of steps, for example harvesting, transport, processing, manufacture, preparation, packaging, one or more storage steps, one or more distribution steps and sale. This lengthy supply chain may end in a food service establishment such as a caterer or in a retailer, typically a supermarket. The food industry tends to use what is known as a Just in Time mode of operation, structuring the food supply chain into hub-satellite networks (DEFRA 2006). In this model manufacturers produce food which is then delivered to distribution centers or hubs. In these centers the food is re-sorted according to requirement, reloaded onto trucks, now as a mixed consignment containing food from many manufacturers, and delivered to a supermarket, wholesaler or cash and carry. From the latter two premises it can be redistributed to independent retailers or caterers. A contamination incident early in this type of food chain, for example at a manufacturer, has the potential to affect a significant number of people. The Sudan dye and melamine contamination of milk examples show this clearly. The trend in the UK in the last 20 years has generally been towards fewer but larger suppliers (DEFRA 2006) providing economies of scale but also potentially greater impact should any contamination incident occur.

Wein and Liu (2005) present a mathematical model which assesses the potential of a deliberate contamination incident in a milk processing factory. This is a good example as the scenario meets many of the criteria listed above for a successful contamination incident – the food is common, popular and inexpensive, it is produced in large batches and has a short shelf life. The method of production and packaging allows thorough mixing of the contaminant. The contaminant chosen for the exercise is botulinum toxin which is certainly hazardous and not noticeable when present in the food, but the approach could be extrapolated to other hazards and production methods (Lui and Wein 2008). The study considers the effect of contamination with various amounts of toxin from 0.1 g to 1 kg. Dilution, cleaning, pasteurization and other parameters are taken into account and the study estimates that between 0 and 568,000 people could be affected, according to the initial contamination and processing parameters. The point of contamination is assumed to be either in a holding tank (farm or processor) or in the transport tanker. However the paper does not address or discuss the plausibility of access by authorized or unauthorized persons at any of these stages. The later paper (Lui and Wein 2008) assumes that a 'terrorist group' could overcome any barriers to access and proceeds accordingly. In fact, where a major milk contamination event has occurred and been investigated (melamine in infant formula) the contaminant was introduced by employees on behalf of the company for economic reasons (Chen 2009), so no barriers needed to be overcome.

4. Controls

Deliberate contamination of the food chain has been identified as a possibility by various authors and government officials e.g. Crutchley et al. (2007); Bénoliel, (2007). Having identified this risk it is necessary to consider its likelihood, what needs to be done to manage it, and, importantly the validity of any controls that might be identified. On page 18 of their document 'Food Security and the UK', the Food Chain Security Group comments on the management of food security. They suggest that having a secure supply does not mean that every risk must be addressed and controlled. There should be balance between the benefits and cost

(DEFRA 2006). This is a perfectly reasonable viewpoint. Even if it was desirable to address every possible risk, however rare or unlikely, it would not be practical to do so. From the examples discussed above some conclusions may be drawn regarding deliberate contamination of food and what appropriate control might be used.

Controlling access to the potential contaminants would be the most ideal way of reducing the risk. Where this is not possible, perhaps because the contaminant is commonly available, or because unregulated stores may exist, the next step would be to control access to the food at vulnerable points. Other safeguards could include testing and detection methods as validation that the food was safe. As an added safety mechanism, good investigation and containment procedures should be in place to ensure any food that might become contaminated is isolated and recalled and that any affected consumers can be identified quickly and treated.

4.1. HACCP (Hazard Analysis Critical Control Points)

With regard to deliberate contamination by pathogens such as *Salmonella* or by toxins such as botulism or nicotine, there are technical barriers to acquiring the contaminants in the first place. As a consequence they are not widely available to anyone wishing to deliberately contaminate the food chain. Even if these barriers can be surmounted, the perpetrator must still access the food chain to introduce the contaminant at an appropriate point. This access may be prevented or at least controlled by correct implementation of good food safety systems. The use of food safety management systems such as HACCP (Hazard Analysis Critical Control Points) has been common in the food industry in the UK for some years and has been a legal requirement under Regulation (EC) No 852/2004 since January 1 2006. HACCP is a system that ensures the production of safe food. This is achieved by proactively identifying hazards at all points in the food process and then implementing measures to control or eliminate those that are critical. Although the system is designed to control hazards that are naturally or accidentally associated with food, when implemented correctly, the system and its control measures may also be effective in reducing and controlling the opportunities for deliberate contamination. HACCP is particularly appropriate for manufacturing premises and was widely used by this sector in the UK even before it became a legal requirement. According to a Local Authority survey reported by the Food Standards Agency in Nov 2001 (just over 5 years before the enactment of Regulation (EC) no 852/2004) 59% of manufacturers were already using a fully documented HACCP system, with a further 27% having some type of hazard analysis system in place (FSA 2002). In a more recent UK survey (Acosta 2005), 96% of the manufacturers in who participated in the study claimed to have implemented these systems.

The control of hazards using a HACCP system can be divided into two approaches. Hazards that are common to all processes such as foreign object contamination from staff, pests, equipment etc. may be controlled using what are commonly called 'pre-requisite systems'. Other hazards, associated specifically with the food/process under consideration, will be controlled by identifying critical control points in the process. At these points a control measure will be implemented to control a specific hazard. An example is the control measure of pasteurization, introduced into a milk bottling plant to reduce the number of mesophilic pathogens to a safe level. The control measures are monitored to ensure effectiveness, records are kept and the control measures verified to confirm the safety of the food.

4.1.1. Manufacturing and Distribution

Both types of control can be extended to address the problem of deliberate contamination in manufacturers. The pre requisite systems typically include controls over staff access to the process and over items that may be introduced to the food premises. This is mainly to control the accidental contamination of clean food by, for example, staff handling dirty material such as waste. However, it also acts as a limitation on the people with access to the food at vulnerable points. Controls on staff apparel, jewellery and other personal items, on eating and smoking while working etc. are actually designed to prevent contamination of the food by accident but have the added advantage of limiting the items that may be brought into the premises. These systems will help to create a barrier for someone trying to bring contaminants into the food premises. This applies whether the contaminants are easily acquired e.g. glass or plastic shards or specialist e.g. toxins or bacteria. It particularly hampers anyone attempting to use a bulky contaminant or needing substantial quantities to overcome dilution issues inherent in contaminating large batches of food. Contamination of raw materials is also a potential risk. Such contamination can have widespread effect as shown by the Sudan dye and melamine contamination incidents. Under a food safety management system, potential contamination of raw materials is a hazard that should be considered thoroughly with the use of reputable suppliers and product specifications as appropriate controls. As all of these systems require supervision and checking, and in some cases screening, the opportunities for deliberate contamination are reduced if such a system is properly implemented in a food premises.

HACCP and food safety management systems are usually the preferred way to control physical contamination such as metal, glass, plastic etc. This is because the systems act to prevent the occurrence of the contaminant in contrast to screening of the finished product where the contamination has occurred and must be identified. Controls would include protecting the food by enclosing it or separating it from sources of contamination. This limits access to the food at vulnerable points. Clearly HACCP and similar systems are not completely fail-safe as evidenced by the examples in Table 1. They do offer an added barrier to someone seeking to deliberately contaminate the food, particularly during manufacture where the principles are widespread and well tested.

Table 1. Examples of accidental physical contamination of foods Nov 2007–April 2009 in the UK

Foodstuff	Contaminant	Date of alert	Reference
Mayonnaise	Metal	April 2009	http://www.food.gov.uk/enforcment/alerts/2009/apr/asdamayo
Scampi	Glass	Feb 2009	http://www.food.gov.uk/enforcment/alerts/2009/feb/kinscampi
Strawberry pencils	Metal	Nov 2008	http://www.food.gov.uk/enforcment/alerts/2008/nov/tescostrawberrypencils
Butter Spread	Rubber	Nov 2008	http://www.food.gov.uk/enforcment/alerts/2008/nov/dairycrest
Potato products	Blue plastic	Oct 2008	http://www.food.gov.uk/enforcment/alerts/2008/oct/croquettes
Apple pies	Hard plastic	August 2008	http://www.food.gov.uk/enforcment/alerts/2008/aug/livwellpies
Crisps	Rubber	August 2008	http://www.food.gov.uk/enforcment/alerts/2008/aug/walkers
Beans	plastic	June 2008	http://www.food.gov.uk/enforcment/alerts/2008/jun/beansausage
Flapjacks	Glass	May 2008	http://www.food.gov.uk/enforcment/alerts/2008/may/flapjack
Cheese spread	Hard plastic	May 2008	http://www.food.gov.uk/enforcment/alerts/2008/may/primula
wine	Glass	April 2008	http://www.food.gov.uk/enforcement/alerts/2008/apr/grandchais
Hot chocolate	Glass	March 2008	http://www.food.gov.uk/enforcement/alerts/2008/mar/hotchoc
Meatballs	Glass	Feb 2008	http://www.food.gov.uk/enforcement/alerts/2008/feb/aldirecallmeatballs
Chocolate	Metal	Feb 2008	http://www.food.gov.uk/enforcement/alerts/2008/feb/mmcchoc
Sauce	Glass	Jan 2008	http://www.food.gov.uk/enforcement/alerts/2008/jan/hollandaise
Naan	Glass	Dec 2007	http://www.food.gov.uk/news/newsarchive/2007/dec/naanrecall
Mince pies	Plastic	Dec 2007	http://www.food.gov.uk/enforcement/alerts/2007/dec/waitrose
Mince pies	Rubber	Dec 2007	http://www.food.gov.uk/enforcement/alerts/2007/dec/kipling
Biscuits	Metal	Nov 2007	http://www.food.gov.uk/news/newsarchive/2007/nov/bourbonupdate
Cheesecake	Metal wire	Nov 2007	http://www.food.gov.uk/enforcement/alerts/2007/nov/cheesecake

Many foods also require product/process specific controls to make them safe (e.g. the pasteurization mentioned above). These control measures will be designed to address the particular hazards associated with that product – for example, the thorough cooking of minced meat to eliminate pathogens such as *Salmonella* and *E coli* O157 or use of specific equipment to prevent re-contamination of cooked product. While these control measures have been designed to reduce or eliminate accidental contamination, the control will clearly also extend to any deliberate contaminants having the same characteristics, thereby reducing further the success for that type of tampering.

The extended food chain common in the UK (Hub-Satellite as mentioned above) offers points of contact with the food (i.e. vulnerability) outside the manufacturer and retailer. HACCP systems are required throughout the food chain, including transport and delivery. The control measures may differ from those implemented by manufacturers but access to the food should still be controlled. Most premises will also include inspection of deliveries in their food safety management system. Such inspections, together with further record keeping, are designed to ensure that the product is of the standard required but can also confirm that the packaging is intact and that no apparent tampering has occurred at a previous step. The use of tamperproof packaging is now widespread in the U.K., providing a barrier to deliberate contamination and a warning indicator if such contamination were to occur. In a good HACCP system, isolation of the product and appropriate investigation procedures usually form part of the corrective action should a delivery fail to meet pre-determined criteria, such as intact packaging.

4.1.2. Retail

In some of the deliberate contamination events described above, the perpetrators accessed the food at retail. Clearly the control measure described for manufacturers and distributors – that of limiting access to the foodstuffs will not be appropriate in this case as the public must be given access to the food at point of sale. However other control measures may be used to limit the opportunity for deliberate contamination. Many retailers also have good supervision of the food on display to consumers. These systems are generally to prevent theft and vandalism but if diligently implemented will further inhibit persons attempting to deliberately contaminate food. In the urine spraying example given above (BBC 2009), the event was observed by both staff and customers. As a consequence the contaminated food was isolated, the perpetrator apprehended and prosecuted and the public protected.

Another important point arises from the example of the blackmailer who threatened to put caustic soda in yogurt (BBC 2008). It is not necessary for someone to actually contaminate food in order to cause a disruption to the food chain. A hoaxer claiming to have done so can also have this effect. The use of tamperproof packaging wherever possible will offer protection against hoaxes of this sort, providing protection throughout the food chain. An advantage of a good food safety management system is that the monitoring, supervision and documentation required can help negate bogus claims or at least quickly contain the relevant food for testing and examination.

It should be emphasized that HACCP plans must be correctly and rigorously implemented in order to be effective against any form of food safety hazard, whether accidental or deliberate. Where HACCP is implemented incorrectly or haphazardly, it will not ensure food safety. The outbreak of *E. coli* O157 in South Wales in 2005 which affected 44 schools, resulting in 157 cases and tragically, the death of a small child (Pennington 2009), is a clear example of this. The food business that caused the outbreak had an inadequate HACCP plan with missing processes and unsatisfactory controls. Some basic pre-requisite systems were not implemented. As a consequence contaminated cooked meat was supplied to schools and eaten by the school children with devastating effect.

4.2. Detection

If a contaminant is introduced into a foodstuff, consumers may be affected unless it is detected prior to sale/ingestion. Wein and Lui (2005) discuss options for post processing detection of the botulinum toxin in their paper. While end product testing has been and is used to some extent in the food in the food industry, it is generally considered to be an unsatisfactory method for routine identification of food safety hazards, especially for microbiological hazards. There are several reasons for this, the main being that it is not entirely effective. Mitchell (2000) provides a simple to understand explanation. He describes the theoretical example of a production line where the contamination rate is 20% of the final product. By sampling 3% of the final product (a sampling rate substantially higher than most manufacturers would be comfortable with) he calculates the probability of collecting a contaminated sample to be less than 50:50. There will also be limitations on the microbiological techniques used which will further reduce the chances of accurately identifying the contamination. End product testing cannot therefore be said to be reliable as a routine method for detecting contamination. As it is also very expensive, it would be impractical to use it to indicate deliberate contamination events, since it would be impossible to predict which organism, toxin or chemical might be used.

Detection of physical contaminants in foods also has variable success. While metal detection is routinely carried out in many manufacturers, contamination with glass and plastic is much more difficult to identify. Methods using X-ray or Ultrasound are subjects of research and consideration and where techniques exist and are practical, the food industry adopt them to prevent the type of accidental contamination described in Table 1. These obviously would assist in also detecting deliberate contamination. However at present the industry tends to rely on preventative measures to control foreign object contamination rather than end product screening as the technical and financial barriers to such screening are still significant and the efficacy is limited. End product testing or screening in a HACCP plan tends to be used as verification that the preventative (control) measures are working at critical control points, rather than as a control measure.

4.3. Investigation and Containment

Clearly the best control for contamination, whether deliberate or accidental is to prevent the incident happening. However, some contaminated food does still reach the consumer and cause illness. This is evidenced by the reports of accidentally caused food borne illness that have been published and incidents such as those referred to in Table 1. Correctly implemented Food Safety Management Systems are considered to be very effective in reducing and controlling food safety hazards but, like any system, will not provide perfect security in all circumstances.

The impact of any incident that does occur can be limited with good containment procedures. Correct investigation of the index case can lead to preventative measures being instigated which limit the potential morbidity and mortality. Such procedures already exist in many countries and are used routinely to deal with accidentally contaminated food which has found its way to the consumer. These include notification and central processing of information such as occurs on Pulsenet (http://www.pulsenetinternational.org/default.asp) together with formal investigations by specialists (Anon 1994). Effective communication, product traceability and recall procedures are also needed. The Rapid Alert System for Food and Feed (http://ec.European Communityropa.European Community/food/food/rapidalert/archive_en.htm) developed under Regulation (EC) No 178/200 links the member states of the European Community plus Norway, Iceland, Lichtenstein and Switzerland in a network designed to share information about food safety hazards. Competent authorities in the participating states, for example the Food Standards Agency in the UK, cascade information as required to protect the public by containing incidents of contamination. The World Health Organisation and Food and Agriculture Organisation set up INFOSAN (International Food Safety Authorities Network) in 2004 to facilitate the international exchange of food safety information. Since its inception it has gathered data on a number of food safety matters and been instrumental in disseminating information to participating countries, allowing a more rapid containment and protection of public health that would have been the case without it (http://www.who.int/foodsafety/fs_management/infosan/en/).

4.4. Traceability and Recall

Regulation (EC) No 178/2002 applies to all countries in the European Union. Articles 18 and 19 impose obligations on food business operators to implement procedures to be used in the case of a contamination incident. Under Article 19 a food business operator (FBO) must inform the food authority (competent authority) of any food which fails to meet food safety requirements or is injurious to health. The FBO must also institute or participate in recalling the suspect food. In order for proper containment to occur, all food businesses must have good traceability as required by Article 18 of the same regulation. In this way food which may be contaminated can be traced throughout the food chain and removed as necessary. National and international Food Alert systems as described above allow any contamination to be notified widely and contained quickly. Food which is imported into the European Union must also comply with the relevant food legislation. This places a responsibility on the importers to ensure that their suppliers have good food safety management systems or other such controls as may be required, thereby enhancing the security of the food chain beyond the boundaries of member states. Traceability systems, the use of INFOSAN and participation of all countries, including China, during the melamine – milk incident helped to contain the problem and also identified areas for improvement in the management of any future incidents (Gossner et al. 2009).

The example of the contaminated oil sold in Spain in 1981 highlights the dangers of buying food from unregulated operators. Registration and/or licensing of food businesses permits better control of food safety hazards by allowing inspection and sanction procedures to be used in businesses selling and supplying food. Rogue traders can be more easily identified and prohibited. It also ensures that when containment is required, all outlets can be covered. Using properly regulated suppliers is essential as shown by the Sudan dye contamination. Appropriate inspection procedures are also necessary even when food businesses are implementing food safety systems. They act as added checks and controls. Chen (2009) reports that in China at the time of the melamine – milk contamination incident, no single government authority was responsible for the control of private milk collection stations. Such a situation of shared responsibility may lead to gaps in control of food safety unless good communication and clear lines of responsibility are established. In addition, the relevant Chinese control authority – Administration of Quality Supervision and Quarantine (AQSIQ) had instituted an exemption scheme since 2000. This applied to businesses with good records and self checking systems and was designed to relieve the burdens on businesses (Fan 2008). This exemption could be indicated on the product label and was used as a marketing advertisement by some participating companies. The main supplier of the contaminated infant formula was participating in this scheme at the time of the incident (D. Zhong 2010).

Deliberate contamination incidents of the food chain are quite rare. From the published evidence it appears that this is due to a combination of reasons. First it is difficult to identify and acquire appropriate contaminants. Even when this is possible, inserting them into the food chain can be problematic, especially if good management controls are in place. On the occasions when these systems can be breached and the contaminants inserted into the food, the effect may often be incapacitating, expensive or frightening rather than fatal, especially if good traceability, notification and recall systems allow isolation of the foods quickly and effectively. It is impossible to summarize or predict the motives of those who perpetrate deliberate food contamination events, but if widespread mortality is their objective, contamination of food stuff may not be the most effective way of achieving that aim, even if a deadly contaminant like botulinum toxin can be accessed. However, that does not mean that the possibility of deliberate contamination should be ignored. Botulinum toxin can be delivered as an aerosol and is apparently far more dangerous in this format than as a food contaminant (Zilinskas 1970; Bossi et al. 2004). Presumably it was for this purpose that the Iraqi scientists were developing their stores. Aerosolisation remains difficult to achieve (Bigalke and Rummel 2005) and although less effective, contaminating food might seem a more achievable goal, should access to the toxin be possible. Another reason to maintain vigilance against deliberate food contamination is suggested by Peter Beinart's commentary in Time Magazine (Jan 18, 2010). As government authorities develop more extensive control mechanisms to try and protect the public from explosive attacks such as the London Tube Bombings, terrorists may respond by considering other potential targets even if the ultimate impact is less devastating.

Analysis of the reports describing deliberate contamination of food indicates that such incidents are not common and that the motives for the contamination include increased profit, personal gain, revenge, and political control. However, it would seem that systems of food safety management such as HACCP and legal obligations such as traceability and premises registration, while designed to protect consumers from adventitious contamination, can, if correctly and diligently implemented, be used to enhance the security of the food chain from deliberate attempts as well. A Food Security Assurance Standard has been published by Product Authentication International (Anonymous 2006) and this provides advice on expanding or upgrading food safety management systems to ensure deliberate contamination incidents are considered.

The use of HACCP based management systems vary between countries (Panisello et al. 1999; Cates et al. 2001; Yous and Sneed 2003; Maldonado et al. 2005; Bas et al. 2006,). It does not appear to be universal in any country but surveys indicate good take up in some parts of the food chain, especially in larger businesses and the manufacturing sector (Panisello et al. 1999; Maldonado et al. 2005). In Member states of the European Community it is now a legal requirement for food businesses to implement such systems. Countries such as the UK have invested heavily to assist the food industry in this implementation. Although barriers have been identified, especially in small businesses (Gilling et al. 2001), there is some evidence to suggest that an improvement in food hygiene and hazard control results from its correct implementation (Cates et al. 2001; Smith et al. 2002, Acosta et al. 2010). An added bonus would be its value to improve security from deliberate contamination attempts with minimal additional expense. Likewise the requirements for traceability, notification and registration are not universal, but where they exist allow effective containment of contamination incidents. Coupled with good inspection procedures and national emergency response systems, for example as occurred in response to the sarin gas attack in Tokyo (Okumura et al. 1998), these will enhance public protection against deliberate food chain contamination. This approach appears to be supported by the SecuFood project (Setola and De Maggio 2009) although the final report has yet to be published.

5. Conclusion

Analysis of accidental and deliberate food chain contamination events suggests that widespread and effective use of food safety management systems such as HACCP can enhance the security of the food chain. The development of notification systems such as the Rapid Alert System for Food and Feed and legal controls such as registration/licensing of premises (food and laboratory), traceability and recall systems would complement the preventative nature of such Food Safety Management systems and help to minimize both the likelihood and effect of a deliberate contamination incident in a cost effective way.

Acknowledgements

The author wishes to express her appreciation to Detective Inspector P. Dingemans of SO15, Jenny Millward of Birmingham City Council and Amir Anvarian, Dan Zhong and Nikolas Hodges of University of Birmingham for their contributions to this paper.

References

Acosta N (2005) *Perception of UK Food Industries on the Costs and Benefits of HACCP*: Unpublished MSc thesis. Division of Environmental Health and Risk Management. The University of Birmingham, UK

Acosta N, P J Fryer & M P Smith (2010) Evaluation of the effectiveness of PRPs (SFBB) and HACCP in small size food service establishments in Birmingham, UK

Anonymous (1975) Sudan I; IARC monographs ; 8:225

Anonymous (1994) Management of outbreaks of foodborne Illness; Department of Health; UK

Anonymous (2006) Food Industry Security Assurance Standard edition 3; Product Authentication International; UK

Anonymous (2007) Report of the Sudan I Review Panel; Food Standards Agency; July 2007

Arnon S, R Schechter, T Iglesby, D Henderson, J Bartlett, M Ascher, E Eitzen, A Fine, J Hauer, M Layton, S Lillibridge, M Osterholm, T O'Toole, G Parker, T Perl, P Russel, D Swerdlow, K Tonat (2001) Botulinum Toxin as a Biological Weapon. JAMA 285(8):1059–1070

Bas M, Ersun A & Kıvanç G (2006) Implementation of HACCP and prerequisite programs in food businesses in Turkey. Food Control 17(2):118–126

BBC (2005) Food Dye Firm in Previous Scare http://news.bbc.co.uk/1/hi/health/4287509.stm accessed Feb 12 2010

BBC (2008) Jail for Tesco Blackmailer http://news.bbc.co.uk/1/hi/7213151.stm accessed Jan 29, 2010

BBC (2009) Man waged Urine spray campaign http://news.bbc.co.uk/1/hi/england/gloucestershire/7906639.stm accessed Jan 29 2010

Beinart P (2010) What al-Qaeda can't do. Time magazine Jan 18 pp 21

Bénoliel I (2007) EU Defending Food Chain against Bio-Attack; European Institute; http://www.europeaninstitute.org/Spring-2007/eu-defending-food-chain-against-bio-attack/Print.html. Accessed Dec 18, 2009

Biagio Solarino B, F Rosenbaum , B Rießelmann, C Buschmann & M Tsokos (2010) Death due to ingestion of nicotine-containing solution: Case report and review of the literature. Forensic Science International 195 (2010) e19–e22

Bigalke H & A Rummel (2005) Medical aspects of toxin weapons. Toxicology 214:210–220

Blake K & J Harvey (2007) Monitoring of radioactive contamination of sheep in Wales after Chernobyl. Government Veterinary Journal 17(1): 22–27

Bossi P, A Tegnell, A Baka, F Van Look, J Hendriks, A Werner, H Maidhof & G Gouvras (2004) Bichat Guidelines for the Clinical Management of Botulism and Bioterrorism- related Botulism. Eurosurveillance 9(12) http://www.eurosurveillance.org accessed Dec 20, 2009

Boulto M, M Stanbury, D Wade, J Tilde, D Bryan, J Pay & B Eisenga (2003) Nicotine poisoning after ingestion of contaminated ground beef. MMWR May 9, 2003/52(18):413–416; http://www.cdc.gov/mmwr/preview/mmwrhtml/mm5218a3.htm accessed Dec 20, 2009

Buzby J & R Chandran (2003) The Belgian Dioxin Crisis and its Effects on Agricultural Production and Exports; Chapter 8, pp 125–136 in: International Trade and Food Safety: Economic Theory and Case Studies. Agricultural Economic Report no (AER 828); http://www.ers.usda.gov/publications/aer828/aer828j.pdf. Accessed Jan 29, 2010

Carrell S & R Whitacker (2005) Ricin: The plot that never was; http://www.williambowles.info/spysrus/ricin_plot.html. Accessed April 29, 2008

Cates S, D Anderson, S Karns & P Brown (2001) Traditional versus hazard analysis and critical control point-based inspection: results from a poultry slaughter project. Journal of Food Protection 64(6):826–832.

Chen J (2009) What can we learn from the 2008 melamine crisis in China? Biomedical and Environmental Science (22): 109–111

Cowden JM & S Ahmed , M Donaghy, A Riley (2001) Epidemiological investigation of the central Scotland outbreak of *Escherichia coli* O157 infection, November to December 1996. Epidemiology and Infection 126(3):335–341

Crown Prosecution Service (2005) CPS statement on convictions of Kamel Bourgass; CPS press release; http://www.cps.gov.uk/news/pressreleases/archive/2005/121_05.html. Accessed June 6, 2008

Crutchley T, J Rodgers, H Whiteside, M Vanier & T Terndrup (2007) Agroterrorism: Where are we in the ongoing war on terrorism? Journal of Food Protection 70(3):791–804

DEFRA (2006) Food Security and the UK: An Evidence and Analysis Paper; DEFRA; http://www.defra.gov.uk/science/Project_Data/DocumentLibrary/FT0352/FT0352_4705_FRP.doc. Accessed June 16, 2007

DEFRA (2009) Government response to the Foot and Mouth Disease 2007 Review; DEFRA; http://www.defra.gov.uk/foodfarm/farmanimal/diseases/atoz/fmd/documents/anderson-090203.pdf. Accessed Feb 1, 2010

Department of Health (2002) A survey of 184 UK brands (Jan-Dec 2002) for tar, nicotine and carbon monoxide yields in cigarette smoke; http://www.dh.gov.uk/ab/SCOTH/DH_096609. Accessed Jan 29, 2010

Dobson R, S Motlagh, M Quijano, R Cambron,T Baker, A Pullen, B Regg, A Bigalow-Kerm, T Vennard, A Fix, R Reimschuessel, G Overmann, Y Shan & G Daston (2008) Identification and characterization of toxicity of contaminants in pet food leading to an outbreak of renal toxicity in cats and dogs. Toxicological Sciences 106(1): 251–262

Fan L (2007) Emergency public health incident response regulation. In: Public Health Law and Surveillance, 2nd Edition, Beijing, People's Medical Publishing House pp 332–38 [in Chinese, translated by Dan Zhong]

Food Standards Agency (2002) Strategy for Wider implementation of HACCP; http: //www.food.gov.uk/multimedia/pdfs/fsa_01_07_02.pdf. Accessed Jan 29, 2010

Food Standards Agency (2005) Sudan I Timeline; http://www.food.gov.uk/safereating/chemsafe/sudani/sudanitimeline. Accessed Feb 15, 2010

Fowle SE, C Constantine, D Fone & B McCloskey (1996) An epidemiological study after a water contamination incident near Worcester, England in April 1994. Journal of Epidemiology and Community Health 50:18–23

Gessler F (2005) A new scalable method for the purification of botulinum neurotoxin type E. Journal of Biotechnology 119:204–211

Gilling S, E Taylor, K Kane, & J Taylor (2001) Successful Hazard Analysis Critical Control Point Implementation in the United Kingdom: understanding the barriers through the use of a behavioural adherence model. Journal of Food Protection 64(5)710–715

GossnerC, J Schlundt, P Embarek, S Hird, D Lo-Fo-Wong, J Beltran, K Teoh & A Tritscher (2009) The melamine incident: implications for international food and feed safety. Environmental Health Perspectives 117(12)1803–1808

Hagiya K, T Mizutani, S Yasuda & S Kawano (2010) Nicotine poisoning due to intravenous injection of cigarette soakage. Human and Experimental Toxicology 2010 Jan 12. [Epub ahead of print]

JFSSG Press release (1999) Dioxin Contamination of Belgian Products – an update; 10 August 1999; http://archive.food.gov.uk/maff/archive/inf/newsrel/fsa/fsa1599.htm. Accessed Jan 29, 2010

Khan A, D Swerdlow & D Juranek (2001) Precautions against biological and chemical terrorism directed at food and water supplies. Public Health Reports 116: 3–13

Kolavic S A, A Kimura, S L Simons, L Slutsker, S Barth & C E Haley (1997) An outbreak of *Shigella dysenteriae* type 2 among laboratory workers due to intentional food contamination. JAMA 278(5):396–398

Lewander W, H Wine, R Carnevale, J Lindenmayer, E Harvey, C Hall-Walker, L Lambright & E Manzo, (1997) Ingestion of Cigarettes and Cigarette Butts by Children – Rhode Island, January 1994–July 1996. MMWR February 14, 1997 / 46(06):125–128

Lui Y & L MWein (2008) Mathematically assessing the consequences of food terrorism scenarios. Journal of Food Science 73(7):M346–M353

Lu J, J Yang, Z Wang, D Jiang, C Fang & J Yang (2009) Study on migration of melamine from food packaging materials on markets. Biomedical and Environmental Sciences 22(2): 104–108

Maldonado E S, Henson S J, Caswell J A, Leos L, Martinez P A, Aranda G & Cadena J A (2005) Cost-benefit analysis of HACCP implementation in the Mexican meat industry. Food Control 16(4): 375–381

Michino H, K Araki, S Minami, S Takaya, N Sakai, M Miyazaki, A Ono & H Yanagawa (1999) Massive Outbreak of *Escherichia coli* 0157:H7 Infection in Schoolchildren in Sakai City, Japan, Associated with Consumption of White Radish Sprouts. American Journal of Epidemiology 150(8):787–796

Mitchell R (2000) Practical Microbiological Risk Analysis . Chandos Publishing Ltd, Oxford
MP press release July 25, 2005 http://cms.met.police.uk/news/major_operational_announcements/terrorist_attacks/press_conference_25_july_14_50 accessed 29 Jan 2010
Okumura T, N Takasu, S Ishimatsu, Miyanoki, S A Mitsuhashi, K Kumada, K Tanaka & S Hinohara (1996) Report on 640 victims of the Tokyo subway sarin attack. Annals of Emergency Medicine. 28(2):129–135
Okumura T, K Suzuki, A Fukuda, A Kohama, N Takasu, S Ishimatsu & S Hinohara (1998) The Tokyo subway sarin attack: disaster management, Part 3: National and international responses. Academic Emergency Medicine 5(6): 625–628
O'Mahony M, E Mitchell, R J Gilbert, D N Hutchinson, N T Begg, J C Rodhouse & J E Morris(1990) An outbreak of foodborne botulism associated with contaminated hazelnut yoghurt. Epidemiology and Infection 104: 389–395
Orre K (2005) The logistics of food supply following radioactive fallout. Journal of Environmental Radioactivity (83):429–432
Panisello P J, PC Quantick & M Knowles (1999) Towards the implementation of HACCP: results if a UK regional survey. Food Control 10(2):87–98
Pennington H (2009) The Public Enquiry into the September 2005 Outbreak of E coli O157 in South Wales; HMSO; http://wales.gov.uk/ecolidocs/3008707/reporten.pdf?skip=1&lang=en. Accessed August 2009
Phills JA, A J Harrold, GV Whitman & L Perelmutter (1972) Pulmonary infiltrates, asthma and eosinophilia due to *Ascaris Suum* Infestation in man. The New England Journal of Medicine 286(18):965–969
Sanchez-Porro Valades P, M Posada de la Paz, P de Andres Copa, O Gimenez Ribota, I Abaitua Borda (2003) Toxic oil syndrome: Survival in the whole cohort between 1981 and 1995. Journal of Clinical Epidemiology 56:701–708
Setola R & M De Maggio (2009) Security of the Food Supply Chain; Proceedings of the 31st Annual International Conference of the IEEE EMBS: 7061–7064
Smith M, S Hussain & J Millward (2002) Effect of the licensing process on hygiene in retail Butchers' premises in the West Midlands, United Kingdom. Journal of Food Protection 65(9):1428–1432
Solarino B, F Rosenbaum, B Rießelmann, C Buschmann & M Tsokos (2010) Death due to ingestion of nicotine-containing solution: Case report and review of the literature. Forensic Science International 195 (1–3): e19–e22
Straub T M, K Höner zu Bentrup, P Orosz-Coghlan, A Dohnalkova, B Mayer, R A Bartholomew, C O.Valdez, C J Bruckner-Lea, C P Gerba, M Abbaszadegan, & C A Nickerson (2007) In vitro cell culture infectivity: assay for human noroviruses. Emerging Infectious Diseases 13(3): 396–403
Stiborova M, V Martı′nek, H Ry′dlova′, P Hodek & E Frei (2002) Sudan I is a potential carcinogen for humans: Evidence for its metabolic activation and detoxication by human recombinant cytochrome P450 1A1 and liver microsomes. Cancer Research 62:5678–5684
Tabuenca J M (1981) Toxic allergic syndrome caused by ingestion of rapeseed oil contaminated with aniline. Lancet Sept. 12: 567–68
This is Gloucestershire (2009) Supermarket Sprayer: Guilty Verdict; http://www.thisisgloucestershire.co.uk/stroud/news/Urine-spray-case-Jury-considers-verdict/article-741606-detail/article.html. Accessed Jan 15, 2010
Török T J, R V Tauxe, R P Wise, J R Livengood, R Sokolow, S Mauvais, K A Birkness, MR Skeels, J M Horan & L R Foster (1997) A large community outbreak of Salmonellosis caused by intentional contamination of restaurant salad bars. JAMA 278(5): 389–395
Tyan Y, M Yang, S Jong, C Wang & J Shiea (2009)Melamine contamination. Analytical and Bioanalytical Chemistry 395:729–735
Van Larebeke N, A Covaci, P Schepens & L Hens (2002) Food Contamination with poly chlorinated biphenyls and dioxins in Belgium. Effects on the body burden. Journal of Epidemiology and Community Health 56:828–830

Wein L M & Y Liu (2005) Analysing a bioterror attack on the food supply: The case of botulinum toxin in milk. PNAS 102(28):9984–9989

Williams C (2008) Six years for Tesco Blackmailer. EHN 23(5):4

Wu Y, Y Zhao, J Li & Melamine Analysis Group (2009) A survey on occurrence of melamine and its analogues in tainted infant formula in China. Biomedical and Environmental Science (22): 95–99

Youn S & J Sneed (2003) Implementation of HACCP and prerequisite programs in school foodservice. Journal of the American Dietetic Association 103(1):55–60

Legislation

EC Decision 2009/9/EC: COMMISSION DECISION of 8 December 2008 concerning the non-inclusion of nicotine in Annex I to Council Directive 91/414/EEC and the withdrawal of authorisations for plant protection products containing that substance
(notified under document number C(2008) 7714)

EC Decision 1999/449/EC: Commission Decision of 9 July 1999 on protective measures with regard to contamination by dioxins of certain products of animal origin intended for human or animal consumption *Official Journal L 175 , 10/07/1999 P. 0070 – 0082*

EC DIRECTIVE 91/414/EEC: Council Directive of 15 July 1991 concerning the placing of plant protection products on the market(91/414/EEC) *Official Journal* 230, 19/08/1991 P. 1–290)

EC Directive 94/36/EC: European Parliament and Council Directive 93/36/EC of June 30, 1994 on colour for use in foodstuffs; *Official Journal L237* P. 13–29

REGULATION (EC) No 178/2002 OF THE EUROPEAN PARLIAMENT AND OF THE COUNCIL of 28 January2002 laying down the general principles and requirements of food law, establishing the European Food Safety Authority and laying down procedures in matters of food safety *Official Journal L 31, 01/02/2002 P. 0001 – 0024*

REGULATION (EC) No 852/2004 OF THE EUROPEAN PARLIAMENT AND OF THE COUNCIL of 29 April 2004 on the hygiene of foodstuffs *Official Journal of the European Union L 139 25/06/2004 P.3–21*

The Colours in Food Regulations 1995; Statutory Instrument 1995 No. 3124; HMSO

Issues of Food Chain Security and Case Studies in the Czech Army

Ales Komar and Pavlina Vasicka

University of Defence, 65 Kounicova, 662 10 Brno, The Czech Republic

Abstract

Food supply system is fundamental extremely open complex. Global challenge is acknowledged and must be considered because food is important source of existence and can be used as a desirable terrorist vehicle. Raw material and food featured intentional versus accidental contamination. Manifestation of global challenges, aspiration for sustainable development and appearance of terrorism create the new paradigm for threats to food safety and defence management.

1. Theoretical Issues of Food Chain Security

1.1. Definitions

Food Safety – Many people do not think about food safety until a food-related illness affects them or a family member. Food safety deals with an unintentional poisoning of food in daily routine. Threats are real and food borne illnesses for hospitalizations known and death are rare.
Food Security – Access by all people at all times to enough food for an active, healthy life. The term concerns mainly the elimination of hunger.
Food Defence – Protection of supply chain defence against intentional, criminal, sporadic, unknown, plausible/unpredictable threats of food chain from farm to fork.
Food Terrorism is an act or threat of deliberate contamination of food for human consumption with chemical, biological, or radio nuclear agents for the purpose of causing injury or death to civilian populations and/or disrupting social, economic or political stability (World Health Organization 2002).

H. Alpas and B. Cirakoglu (eds.), *Food Chain Security*,
DOI 10.1007/978-90-481-9558-9_4, © Springer Science + Business Media B.V. 2010

1.2. Principles

These are main topic *principles* in spheres of business concerning food subsistence in the military resort:

- Food quality
- Food safety
- Food defence
- Food protection which includes animal protection and crop protection

The food supply system has only partial influence what concerns spheres of business as they are usually apart. Functional requirement of times is achieve in supply chain complexity to cover all spheres as aggregate function.

Principles of *delivery* must respect that the delivery of that food supply is highly dependent on the many facets of the food system and food supply chain. Distribution chain is not a simple system. Actually with globalization the world global food system is created with many strength and significant risk (imported foods, foreign facilities registered, low part usually only 3% of imported food is inspected at the border and borders are porous).

Food Quality/Safety – we understand food-borne illness due to *system failure*
Food Defence – we understand protection against food-borne illness due to *system attack*

Food-borne Illness due to System Failure

Processing/manufacturing systems are designed to prevent accidental contamination, mitigate harm from anticipated contamination (pasteurized milk) and prevent cross contamination in food processing and food service (carcass washing, raw/ cooked tools), prevent human/environmental contamination ("clean room" processing). Contamination results from poor system design or failure.

1.3. Perception

Perception must respect systems thinking that is imperative for complexity. Principles of food quality, safety and defence, protection in processing, production, transportation and distribution are fundamentally same but we must recognize similarities and differences. E.g. symmetrical versus asymmetrical conflicts causes different conditions for application of principles and perception of food chain security.

Food Safety Risk Management

- All hazards approach from farm to consumption
- Identification and prioritization of risks
  - o Probability of the risk
  - o Consequences of system failure

- Implementation of justified interventions
  - o Reduce/eliminate the potential for system failure
  - o Economically viable or regulatory requirements
  - o Repeat analysis/intervention cycle until cost/benefit act acceptable

Food Safety Interventions

- Pathogen inhibition
  - o Vaccines (animals)
  - o Probiotics (animals and environments)
- Pathogen inactivation/elimination
  - o Sanitation systems (cleaners, biocides and processes)
  - o Terminal processing (pasteurization, additives)
- Detection and control
  - o Real time detection to contain contaminated product
  - o Public health surveillance systems
  - o Supply chain management systems for recall

Food Safety Risks in Current Conditions

Preventing system failure when impact is severe and exposure likelihood low, medium or high and when exposure likelihood is low and impact severe, moderate or mild. Other impact and exposure likelihood are covered HACCP.

Food Defence Risks in Current Conditions

Successful system attack preventing in case when impact is severe and exposure likelihood is low, this protection belongs to defence domain.

Intentional food contamination has an old and new concern

- Food as a vehicle for causing illness an ancient military weapon
- Recent food contamination events localized (e.g. in US)
- Contaminations since 2002 include: rat poison, salmonella (Beza and Komar 2005), nicotine sulphate, arsenic, folimate (garden insecticide) and bleach/ ammonia
- Significant food contamination challenges versus food safety concern microbial and chemical agents, agents that are not normally present and complexity of the system means many vulnerabilities

Intelligence on food terrorism has revealed that terrorist groups have a lesser interest in biological materials and are most interested in chemicals such as cyanide salts to contaminate food and water supplies (CIA Testimony to Congress 2002).

2. Military Food System Unique Challenges

- Military a high profile target with food safety events already (vendor recall)
- Deployed military food sources combine shelf stable, sourced foods (MRE's), elimination and control sourced perishable foods, controls fresh Locally sourced foods and elimination locally prepared final food products (Fryc et al. 2005)
- Impact of contamination could be mission failure

The problem are challenges coming in various forms as examples mainly "mother nature", disgruntled employees, violent activist groups, criminals/subversives, international/government supported or directed groups or individuals (terrorists)

Intentional contamination has following target characteristics

- Opportunities for perpetrator access
- Lack of subsequent inactivation treatments
- Large volume and/or maximum mixing
- Rapid, broad distribution
- Rapid consumption
- Disproportionate consumption by high risk populations
- Disproportionate consumption by high target groups

Considerations about intentional contamination stream from concentration of agents is likely higher, types of agents are unexpected, choice of targets for stealth and effect, scale/number of targets for impact and response to outbreak requirements faster.

Challenges for food defence and **risk management** consider all hazards approach for things that can't happen from farm to consumption. Identification and prioritization of risks are based on vulnerability only and economic impact lacks probability function. Implementation of justified interventions likely requires regulation/command or other motivation. It is difficult to prove intervention value.

Challenges for food defence **interventions** are based on pathogen inhibition (vaccines less attractive, probiotics not effective against chemicals/toxins), pathogen inactivation/elimination (biocides/terminal processing often not validated, inactivation of chemicals/toxins very difficult) and detection and control (real time detection of multiple, unanticipated agents, supply chain management must be real time control).

Food defence strategies reduce the potential for catastrophic events, e.g. provide by rendering targets unattractive, rapidly and accurately detecting attacks, responding effectively to minimize consequences, rapid delivery of effective recovery efforts and training new scientists and professionals.

First need is to prioritize where to start. The problem is e.g.; differentiating normal from intentional contamination following appearances, as types of information, reports, situations, types of agents, concentration of agents, choice of targets, scale/number of targets and recognition of illness, cause, response, is helping to solve the problem.

3. Food Defence Strategies

Key issue is where to focus interventions. Eliminating food as a target seems to be unrealistic. Appropriate defence measures are rendering targets unattractive for potential subjects, multiple barriers to effective contamination in military approach (guns, gates and guards) and in "hardened" processes, minimizing impact of an effective contamination as in prevention and agent inactivation via processing/ formulation or rapid, effective public health response.

3.1. Supply Chain Management

In Globalize market with food and raw key tendency must be partnerships and technology to enable supply chain protection because technology is still nascent and partnerships need to be global to avoid risk food chain security. In this time are common best practices and standards shared up/down the supply chain, real time communication across the supply chain and positive control throughout the supply chain. In practice realization lay in army approved suppliers.

3.2. Food System Event Modeling

There is a need for functional models to examine food system vulnerabilities to guide intervention efforts and support training. In this way vulnerability assessments conducted with limited private sector participation is a challenge like that public health response and capability information limited.

It is necessary to start a new approach. Pilot study presentations of nation's food chain security have shown that military resort plays either margin or no role in the governmental system and private sector and state agency partnership.

3.3. Detection

Rapidly and accurately detecting attacks is part of food defence strategies in military. Real time detection of contaminants in food consists of detect to prevent, detect to protect and detect to recover. As current technologies are insufficient and desired performance unachievable the situation needs of key detection strategies that meet the speed, sensitivity and specificity.

3.4. Conclusion Vision

Defending the safety of the food system is responsibility of provisional and veterinary professionals in the military. To keep food chain security (right nutrition and safety) on high level expertise it is possible with military advisors only and through their research and education.

These experts would manage supply chain and share information, e.g. evaluate best military practices and develop standards and conduct online information systems on food chain security.

I daily practice experts will check your own situation as follows preparedness and readiness to respond, capability of recovery, assess/minimize vulnerabilities, consider an all-hazards approach (quality to pandemic), insist on a systems approach (failure vs attack), understand the product, process, distribution, consumer, responsible agencies, and all associated characteristics (Komar et al. 2005; Beza et al. 2006).

4. Czech Army Case Studies' Excerptions

There are no problems with food chain security considering overproduction of the food, high technology and hygiene of production in the Czech Republic, and the production is local with national labour. **Rational nutrition** ensures energetic and biological rate and well-balanced composition of food sorts.

The opening of the Czech market needs implementation of the EU legislation. HACCP has already been implemented. The DoD has stopped earlier controls of food special produced for the Czech Army subsistence. Some diseases are frequently common as a subsequence of consumption culinary finished food (enteritis, Salmonella, Campylobacter). Similar problems are in farther regions out of EU. Food chain security is then solved with military food rations. These are safety but lack of proper bread can cause adversely (Komar et al. 2006a).

4.1. Introduction

Problems with provision of the food products in the military operation and mission include security, diet, psycho-social and economical, technological aspects. Experience from KFOR and ISAF missions is studied mainly from nutritional aspect (Komar et al. 2006b, c):

Security aspect: The provision secure and hygienic safety food.
Psychosocial aspect: The impact to the total well-being (welfare) and efficiency of nutrition. Infringement of the traditional food decreased physical immunity of the person, frustration tolerance.
Diet aspect: Indigenous food or alimentary distress and indigestion – according to the standard STANAG 2937 on nourishment of the soldiers in field should comply with their national catering usage.

The ways of contingent provision:

- Purchase from the local suppliers. Risks: unstable microbiological quality, substandard sensorial quality of the products.

- Purchase from the allied forces, at concert pitch toast bread. Risk: diet and sensorial value.
- Supply from the Czech Republic. Risk: non-productive.

Safety subsistence in good quality is achieved by the traditional food, national chefs, containerised military kitchen, and containerisation of field bakery. The safety subsistence fulfils preserved military rates.

4.2. Food Servings

Czech Army provides soldiers with hot food three times a day and it is necessary to manage supplying troops with food in the field. One option is to use the ready-to-eat food rations as a combat food ration (thereinafter CFR) in the field training with their fast preparation in the field kitchens. Another option is unitized group ration (UGR) that is the ration of ready-to-cook food in larger packing for an organic unit training in the field. Such way of catering ensures safety, hygiene and requires minimum technical equipment for preparing meals as well as minimum number of service personnel and low level of expertise. At present time the use of UGR leads to the consumption of safe, indigenous food. It is very important in a global world with free food market (Komar and Fry 2007).

Characteristics of food servings are developed for reasons of providing a full daily nutritious meal to individuals, who cannot obtain food by any other standard available means. The CFR and UGR have to meet the basic food ration (BFR) energy requirements increased by the field bonus value that the BFR energy value is 14,560 kJ and the field food bonus is plus 1,700 kJ and provide nutrition for an individual for 24 h. They can be used not only by the military during battle readiness, or war-like situations, but also in natural disasters, or in other operations of the integrated rescue system not only in the Czech Republic. The availability of a range of variants enables its repeated use that is limited to 30 days maximum according to STANAG 2937 and to avoid symptoms of food fatigue in case of long-time consumption.

Minimum durability of finished dishes is at least 26 months and they are packed in small laminated aluminium containers. They are prepared by heating in packages in water baths or an extra set for the heating of food. Other requirements for the military rations are minimal demands on service personnel and consumption of drinking water, consumer packaging for standardized number of personnel (e.g. 1, 25, 50, 100), variable assortment of meals for 7 days, functional package and compliance with laws on packages and wastes and easy preparation in field kitchens for larger number of personnel during an armed conflict, support of units in missions abroad, in crisis situations and on exercises.

Finished dishes mainly in glass and metal plate cans for four-member family used to be a traditional product in former Czechoslovakia and it would be good to restore this tradition for the Army needs.

4.3. Bakery

The bakery products and primarily bread are essential for healthy nutrition and current alimentary and military practice. Bread is still considered to be essential and irreplaceable food in soldiers' nourishment within the Army of the Czech Republic. There are several advantages of bakery products, such as consumption without heat treatment, no danger of overeating, eating without limitation – ad libitum, in order to gain satiety, energy and necessary nutrients and possible role of single alternative food. Bakery products are important side-dish in military combat rations. Bread ration is 400 g a day for one soldier in the field.

Traditional food in realistic combat conditions is important part of national awareness. STANAG 2937 Standardization of Combat Rations declares necessity of food subsistence corresponding to national usage during operations led NATO.

Durable bread will be used as a supplement to food rations. Prospectively it is assumed that durable bread will replace other durable products, such as biscuits and crackers. This product is also in demand in our units abroad. Our soldiers have their eating habits and refuse some products of foreign origin due to their e.g. sensorial properties and hygienic reasons. It is also necessary to supply safe and healthy food for soldiers in foreign missions. We are certainly aware of the fact that the Army fulfils tasks in foreign missions and emergency situations, when the bakery products cannot be bought any time. Therefore it is effective and suitable to introduce durable bread in these specific conditions (Sedivy et al. 2005).

Farther-out missions demand production of bread directly in the place. Advantage is flexible technology which enables production of bakery indigenous products for joint operations.

4.4. A New Technology

Premix allows producing **functional foodstuffs** with special properties in accordance with mission tasks. Substantial bread with prolonged durability and protective factors (higher rate of fibre, addition of pre-biotic micro-organisms, etc.) will be supplied e.g. to the soldiers on several-day patrols without possibility of continuous supply. Raw materials suitable for the composition of 100% premix have been tested. They are prepared by the mixing of powdery ingredients and biologically inactive ingredients. Such premix allows product to be prepared by adding dry yeast and water only. The basic raw materials were flours, cereals, additives, spice, spreading and improvers.

The latest demands of troops for subsistence can be solved by implementing the innovated baking technology. **New and modern technologies** of bread and pastry production simplify and shorten technological time, remove technological risks of production and with improvers both efficiently attain standard quality and broaden assortment of bakery products. Innovated technology of bakery products ensures: healthy, hygiene safe products, high-quality and standard products, minimized production costs, production rationalization, production in a container field bakery, development of products with protective effects and ability to meet all challenges.

The innovated baking technology can be processed in a smart **modern field bakery** (containerized or mobile).

Guarantee safety food chain for supplying the forces is really easy with the help of premixes, their transportation, packaging, storage and own production – simply, undemanding, germs-free, fresh and indigenous bread, effective and efficient.

4.5. Advanced Management

IT and Enterprise Resource Planning specialist CSB – System can facilitate food chain security whole process monitoring. System is represented across Europe, the United States, and China. Computer software business consulting system is perfectly attuned to meeting the business safety requirement of the batch oriented process relating to all processes. In addition to the materials flow and the information flow that accompanies it, all processes and telemetric requirements are visualized by a single software solution. CSB – System presents Coverage Planning Management in an effort to assist with the 100% availability of products at all times (Vasicka et al. 2007).

References

Beza T, Komar A (2005) The Incidence of Alimentary Diseases in the Army of the Czech Republic: Special Focusing on Campylobacteriosis. RDA Proceedings. http://www. militaryfood.org. Accessed 10 December 2005.

Fryc J, Komar A and Sroll B (2005) Development of Ready-to-Eat Food Rations. CATCON 05.

Komar A, Fryc J and Sedivy M (2005) Main Objectives of Military Food Development in the Czech Army. RDA Proceedings. http://www. militaryfood.org. Accessed 15 August 2005.

Beza T, Komar A, Novotny R and Musil M (2006) Food Safety Ways in the Army of the Czech Republic Research Projects. IAFP's. http:// www.foodprotection.org. 15 November 2006.

Komar A, Novotny R and Sroll (2006a) Food Supply to the Army of the Czech Republic. CATCON 06.

Komar A, Beza T, Musil M and Vasicka P (2006b) Nutritional Level Survey of the Czech Army During NATO Led Military Activities. RDA Proceedings. http://www.militaryfood.org. 2 December 2006.

Komar A, Beza T, Musil M and Vasicka P (2006c) Survey of Soldiers′ Nutritional Level During Foreign Mission. RDA Report 59:33–35.

Komar A and Fryc J (2007) Unitized Group Ration (UGR) Contracting Topics Investigation. RDA Proceedings. http://www.militaryfood.org. 1 December 2007.

Sedivy M, Komar A, Sroll B and Hrabe J (2005) Development and Use of Durable Bread for the Needs of the Army of the Czech Republic. CATCON 05.

Vasicka P, Dvorak J and Komar A (2007) Advanced techniques of logistics support. University of Defence prints.

Food Chain Security in Romania

Liviu-Daniel Galatchi and Diana-Lacramioara Mihalache

Ovidius University of Constanta, Constanta-4, Romania

Abstract

During the last years, Romania, as a member country of the European Union, has achieved much progress in the transposition of the Community in the field of foodstuffs. According to the commitments assumed through the position documents during the negotiation process and in order to approach in a unitary way in the field of food safety, the legislative approaches were initiated to promote and adopt a legislative document which may lead to the establishment and organization of a structure corresponding to the European model. In this case, a questionnaire was considered suitable to get a brief and objective data about the food safety, in Romania and other European countries. The questionnaire has been translated and adapted to facilitate the work of the participants, assuring that none of the questions is altered. Most of the results were obtained from the processing sector. The questionnaire was a suitable and useful method for achieving knowledge concerning food safety. It showed that the last years brought an important evolution regarding food industry and its safety around Europe, but still there are many actions that need to be taken.

1. Introduction

Food chain security is a real issue nowadays, never in human history has existed such an abundance of food stuff in most countries like now. And still, let's not forget that not everything goes properly.

Lately, many governmental and non-governmental organizations involved themselves in the matter of food security. Even if it's about ensuring the food stuff from processing to consumption or it's about ensuring the entire food chain against possible terrorist attacks, the important fact is that today there are many specialists that are discussing and searching proper solution for the future concerning everyone's alimentation (World-Watch Institute 2008).

A study with the same concern, appertaining to the Committee on the Challenges of Modern Society (CCMS), and supported by NATO and the European Science Foundation (ESF) is the Pilot Study called briefly "Food Chain Security". Its objective is to study the safety of food stuffs in face of a defective handling and

H. Alpas and B. Cirakoglu (eds.), *Food Chain Security*,
DOI 10.1007/978-90-481-9558-9_5, © Springer Science + Business Media B.V. 2010

also related to possible terrorist attacks. Also, the study searches for response measures in reducing future risks and mitigates the consequences of these threats to the food system.

The study started in Turkey in the year of 2003 and since 2008, seven meetings took place to which countries like Algeria, Belgium, Czech Republic, Finland, Germany, Hungary, Italy, Netherlands, Portugal, Romania, former Yugoslav Republic of Macedonia, Moldova, Russian Federation, Spain, Turkey, United Kingdom and USA, participated.

Romanian contribution began with the sixth meeting in Lisbon (Portugal) that was held in January 2007 and continued in the seventh meeting in October 2007 from Birmingham (UK).

2. Elements Concerning the Food Production and Consumption in Romania

Both production and consumption for a country, represents two statistical indicators that can help on knowing the state of food security. After analyzing the evolution of Romania's agricultural production, imports and exports as well as food consumption during 1990–2007, but also through comparison with other countries in the EU, several conclusions can be highlighted.

During the period 1990–2007 the major increase of the cultivated area in Romania was observed at wheat (in 2006 about 73% more), and the least cultivated were the vegetables with 0.44% from the total cultivated area in 2006. As for the evolution of the vegetable production, in 2006 the quantity of cereals has increased (with 2,030 t more than in 1990) but the production of sugar beet has diminished (with 2,605 t less than in 1990). Regarding the animal production, during the years 1990 and 2004 the animal effectives have decreased significantly, some continued to decrease until 2006 (for example cattle – with 700,000 heads less than in 1994). A better situation can be noticed at birds effective (15% increase in 2006 than in 1998), goats and bees (with 21% and 36% in 2006 than in 1999) (Romanian National Institute of Statistics 2005, 2006).

Exports in Romania during the years 1990 and 2006 have increased cu 26.8% in 2007 unto 2005 (living animals have the most important weigh 20.9%) and imports recorded a grown of only 19.9% (the most important was at pork). The fact that Romania has transformed into a truly importer shows that the country is not able to ensure its own food production to support consumption and so, the food security (Romanian National Institute of Statistics 2007).

In what concerns the food consumption on habitant, it has improved during the last years, increasing at almost all products (the major increase was at milk with 25% more). The analysis of the food consumption in calories shows a growth of quality products in consumption but their quantity still remains insufficient. Also, the consumption in proteins has increased with 112.72 g day^{-1} person^{-1} during 1996–2006, most of them had animal provenance (62.87 g). In the structure of the lipids consumption the most important weight has the animal processing (56.56 g day^{-1} person^{-1} from the total consumption of 104.54% in 2006) but the growth of the food consumption in lipids between 1990 and 2006 is more due to the growth of the lipids with vegetable provenance (Guvernul României 2008, Romanian National Institute of Statistics 2008).

In comparison with other countries, in Romania became clearly, mostly the differences in the quality structure of consumption (a French eats with 38.2% more qualitative than a Romanian and at global scale, Romanians eat with 27% more qualitative).

3. Work Methods

"Food Chain Security" searches to study the safety of food stuffs and to propose different measures for reducing the risks that may affect the food system. The prevention of problems can't be practical there fore the existence of a risk management is needed. In analyzing food security and also for a statistical evaluation, a survey was proposed to be done in all the participant countries (Figure 1) based on a questionnaire composed of 15 questions plus the dates of the companies who completed it (Table 1).

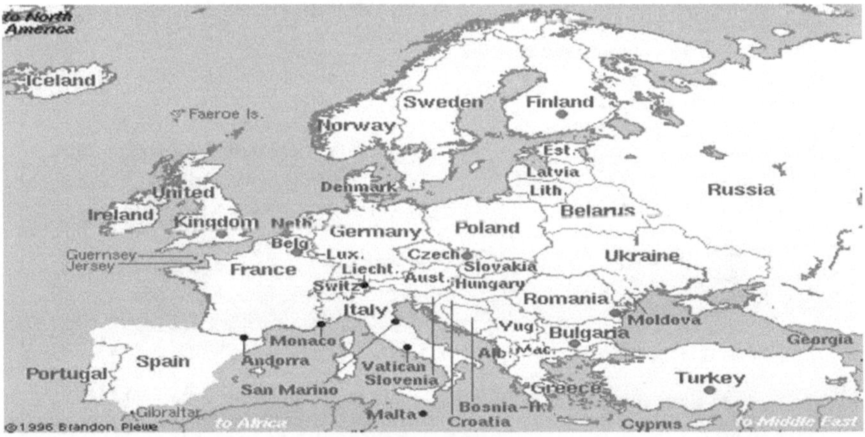

Figure 1. Participant countries in PS "Food Chain Security" (yellow) and countries that did the survey about food security with the help of the questionnaire (red)

The existing questions are grouped in three categories:

- Question 1, 12 and 13 searches to find out the point of view of the participant companies, what are the possible risks for their products, the vulnerable points that could endanger the food security and their opinion related to the possible threat of GMOs.
- Questions 2–5, 10, 14 and 15 are about the measures that the participant countries take in risk management, the measures against possible terrorist attacks and the monitoring of food security and raw materials.
- Questions 6–9 share information about the companies that agreed on completing the questionnaire, about their sector of activity in the food chain (processing, transport etc.) and about the types of products they handle.

Table 1. Questionnaire

Box. 1

Participants information

Identification questions:

1) Please identify which sector of food chain you work in.

Agriculture
Transport
Production
Depoziting
Distribution

Definition:

- Agriculture: raw material concerning farms, feed industry, farming.
- Transportation 1: from feed producer to farm
- Transportation 2: from farm to the processing
- Transportation 3: from processing to distributor
- Processing: all steps of transformation of raw material into consumable products
- Storage: silos, cool-houses.
- Distribution: from distributor to retail shop

2) What kind of products do you produce?
1. Milky
2. Meat
3. Cereals
4. Vegetables
5. Fruits
6. Drinks
7. Others (please specify) ………………

Questionnaire

1) Which of the followings would be the most important risks for your product:
a) chemicals,
b) bacteria/virus
c) radiation
d) energy crisis
e) water crisis
f) raw material
g) GMO
h) Others...(please, specify)

2) Do you have internal discussion in the company about terrorist attacks?

3) Do you regularly check your employee records?

4) Do you have a food safety monitoring program?

5) Do you carry out any one of this inspection before you accept the raw materials:
a) chemicals
b) microbiological
c) radioactivity

6) About your suppliers:
a) domestic
b) international

7) About your company's holders.
a) Multinational
b) National

8) Who is in charge of the food distribution?
a) Yourselves
b) Your contractor

9) Who is in charge of food storage?
a) Yourselves
b) Your contractor

10) How quick do you think that you can respond to a food safety problem?
a) immediately
b) within a few hours
c) a day
d) can not tell

11) Do you have crisis management system?
a) yes
b) no

12) Which of the followings is more vulnerable from the point of food safety?
a) Raw material
b) Processing systems
c) Storage systems
d) Distribution systems
e) Retail shops

13) Do you think that GMO is a safety concern in your production system?
a) Yes
b) No
c) No idea

14) Do you have a recall system for the products?
a) Yes
b) No

15) Do you have tested your recall system?
a) Yes
b) No

Thank you for being involved!

First, the questionnaire was translated and adapted in each country's language. Further they were multiplied and distributed to several economic agents from the food sector. Their number varied from a country to another. In Romania, 30 companies received the questionnaire but only 27 accepted to complete them and respond to the 15 existing questions.

The data obtained from them was gathered and the answers were quantified after a specific model, the same for every participant country. For example, at the first question, when the answers were introduced, both the box corresponding to the sector of the food chain in which the respondent worked (agriculture, transport, processing etc.) and the box that showed what kind of products did the respondent handle (dairy, meat, vegetables etc.). For those that worked in more than one sector, for example in processing and storage, the same answer was entered in two boxes (one for processing and one for storage). The same happened when a respondent gave more than one answer at a single question. For example, at question 5 (inspection carried out before accepting the raw materials) many responded that they do a chemical and also a microbiological inspection and the answer were entered in the two boxes corresponding to the two inspections (Table 2).

Finally, after gathering the data and entering it in the forms, the information then is interpreted by analysts. Figure 2 represents the steps presented previously.

Table 2. Example of the data form with the answers

Q 5	Agriculture	Transport	Processing	Storage	Distribution
Y	0	6	16	5	2
Y(a)	0	3	10	4	1
Y(b)	0	4	11	3	1
Y(c)	0	0	0	0	0
No	0	0	0	0	0

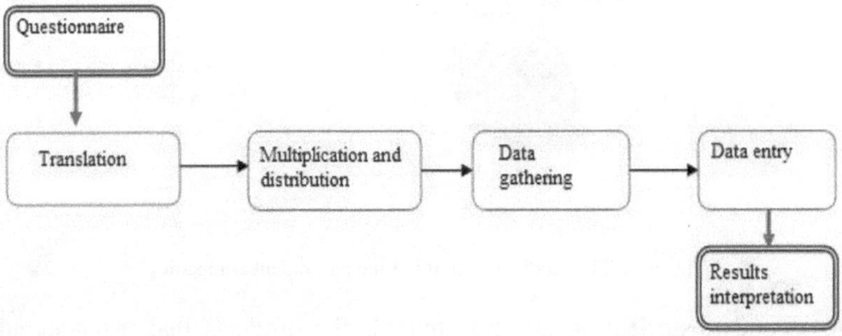

Figure 2. Steps followed for obtaining information about food security through the questionnaire

4. Participant Romanian Companies from the Food Sector

The first part of the questionnaire contains two identification questions for the participant company. They search to identify the sector of the food chain where the company works namely, agriculture, transport, processing, storage and distribution.

For a better understanding these terms were defined as following:

- Agriculture: raw material concerning farms, feed industry, farming
- Transportation 1: from feed producer to farm
- Transportation 2: from farm to the processing
- Transportation 3: from processing to distributor
- Processing: all steps of transformation of raw material into consumable products
- Storage: silos, cool-houses
- Distribution: from distributor to retail shop

The majority of the respondents from Romania are processors (56%); some are also involved in storage or distribution. All the sectors were covered except agriculture (Figure 3):

- 22% worked in transport
- 7% in storage
- 7% in processing and storage
- 4% in storage and distribution
- 4% in distribution

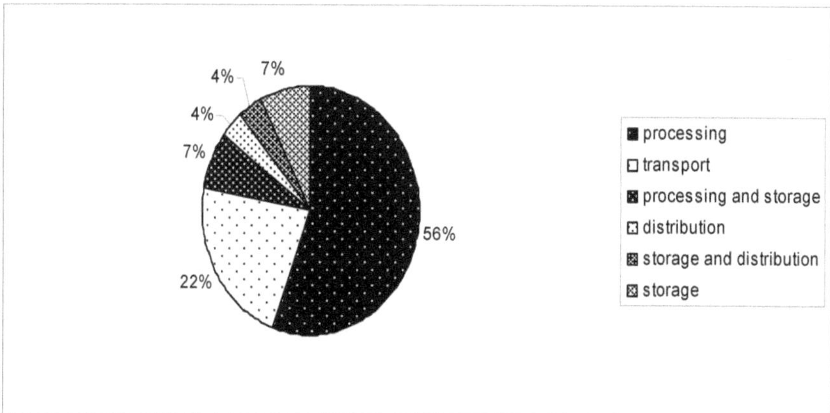

Figure 3. The sector of activity of the participant companies

The second identification question regards the products that the respondents handle. They could choose between dairy, meat, cereals, vegetables, fruits, beverages or other products. Most of the answers were for meat (47%). Because of a typing error and some problems in translation none specified for "other" which exactly were those products although many answers were received here (30%). Dairies also were well represented (11%) followed by fruits and cereals, both with 7% each and cereals, the category with only 4% of the answers (Figure 4).

Another question regarding the participant companies is question 6, about the suppliers. According to the answers, more than a half (55%) had national suppliers, 15% prefers to buy international and 15% both national and international suppliers.

Question 7, national or multinational company, showed that 66% of the participant companies were working at a national level and 34% were multinational companies (only one company marked both answers).

Question 8: "*Who is in charge of the food distribution?*" The majority choused to leave this responsibility to a contractor (52%), only 33% assumed it and other 15% share it with their contractor (Figure 5).

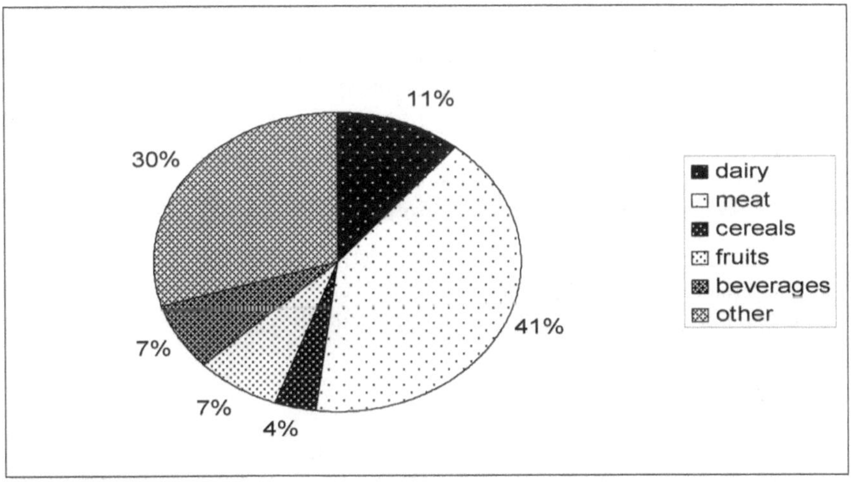

Figure 4. The sector of the food chain and the types of products corresponding to the participant countries from Romania

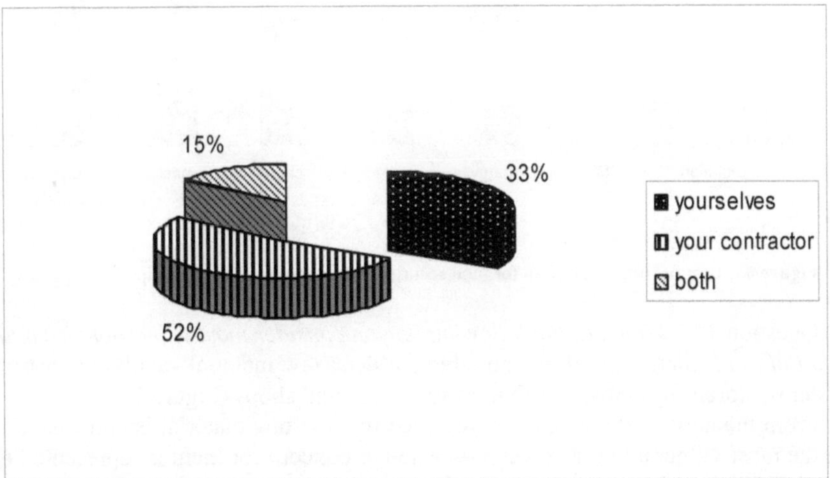

Figure 5. Responsible with food distribution

When it comes to storage (question 9: "*Who is in charge of food storage?*") some rather do it personally (59%) meanwhile others have a contractor in charge with it (19%) and 22% share the responsibility.

5. Risks and Vulnerable Points for Food Security in Romania

One of the most important questions of the questionnaire for food security is the first one: "*Which of the following would you consider as important risks for your product?*"

As it can be observed from Figure 6 that shows only the potential risks in the processing sector (which is the most represented), water crisis is considered to be the biggest risk for food security by the processors (59%). This is not surprising knowing the meteorological problems that existed in Romania in the last years. Another important risk, like the graphic shows, is about the raw material (52%), the other answers reflecting the following situation:

- Chemicals: 41% of processors are considering them an important risk.
- Bacteria/viral contamination: 44%.
- Radiation, genetically modified organisms (GMO) and other: 0%.
- Energy crisis: 41%.

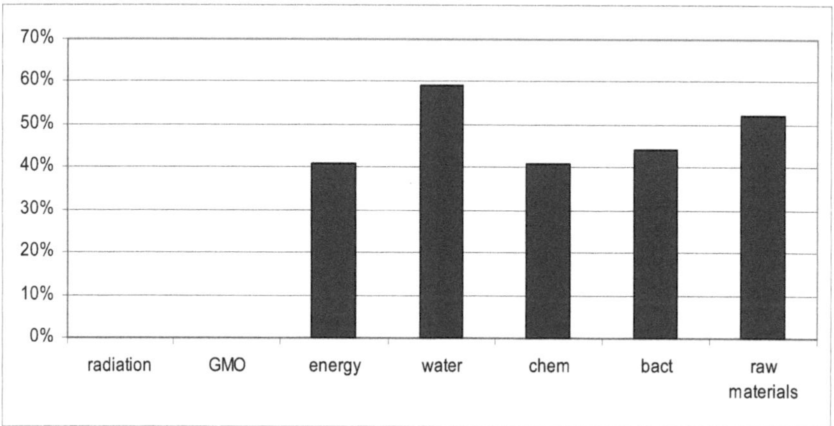

Figure 6. The most important risks for food security in the opinion of the participant companies

Question 12: "*Which of the following do you consider most vulnerable from the point of food safety?*", with the possible answers: raw material supply, processing systems, storage systems, distribution systems, retail shops (Figure 7).

From the above graphic can be observed that the raw material is considered to be the most vulnerable for processors. Another concern for them is represented by storage and distribution systems. On the other hand, raw material doesn't seem to preoccupy those that work in distribution or transport.

Question 13: "*Do you think that Genetically Modified Organisms would be a safety concern in your processing system?*" As we can notice from Figure 8, more than half of the participants don't consider them to be a threat, 11% are feeling threaten and 33% don't know yet.

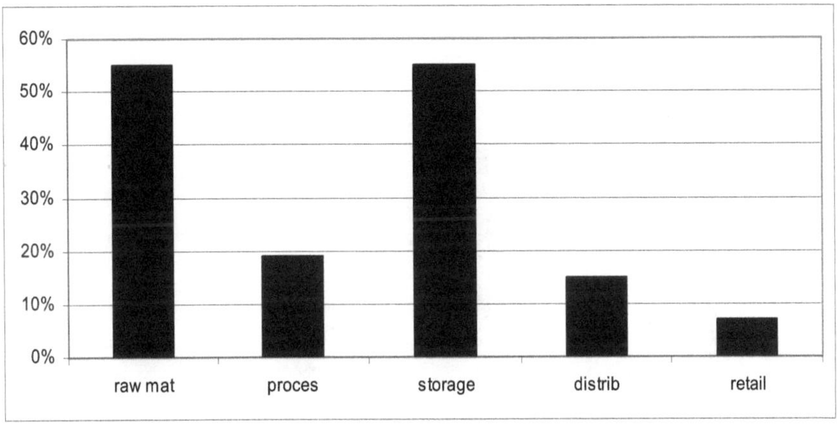

Figure 7. Vulnerable points for food security in the opinion of the participant countries from Constanta (Romania)

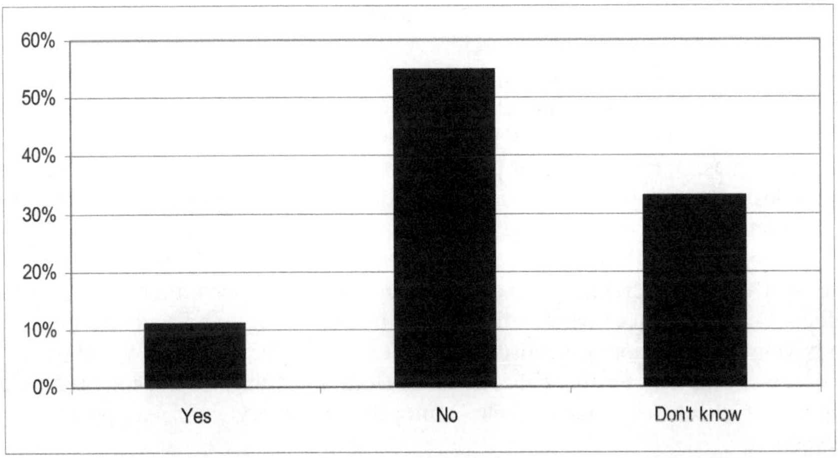

Figure 8. How dangerous GMOs are in the opinion of the participant companies

6. Risk Management

None of the participant companies from Romania have meetings or formal discussion about terrorist attacks (question 2).

At question 3 "*Do you make any security checks on your employees? If yes, when do you do this?*" all answers were affirmative, but only 70% make this checks before employing (Figure 9).

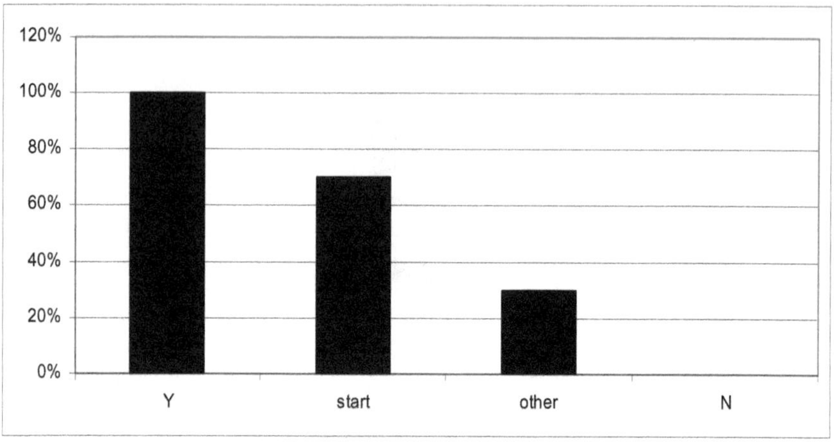

Figure 9. The percentage of the participant companies from Constanta that makes security checks on their employees and when exactly takes place

There were some problems with this question because of the translation and also of its later reformulation. Therefore, two types of questionnaire existed at the same time that differed only at questions 3 and 4, about the monitoring of food security. Although initially incomplete questionnaires were distributed, in the end we managed to obtain answers for questions 2 and 3, corresponding with the complete questionnaire. Another problem was about its meaning and if some how it was lost in translation. Question 3 is about checking the employees in the idea of preventing any terrorist attack.

The concern for these threats grew after the September 11 attacks in US. An attack of this kind targeting one state's alimentation can have much more worse effects. Alimentation is what defines us as human beings, and without food, the body wouldn't function. Likewise a microbiological attack on food would be hard to discover and with terrible consequences. Because of this, is very important that companies check periodically their employees and every time new persons are brought in.

Another issue that makes alimentation so vulnerable in face of a terrorist attack is also the complexity of the food chain. For example, in the case of a chemical contamination or a microbiological one, the identification of the source can be done after verifying the entire food chain, from raw material to processing and consumption, which is quite difficult.

From the answers received at question 4: "*Do you have a food safety monitoring program?*" resulted that only 89% of the companies have such a program, some HACCP (Hazard Analysis and Critical Control Point), others didn't specified (Figure 10).

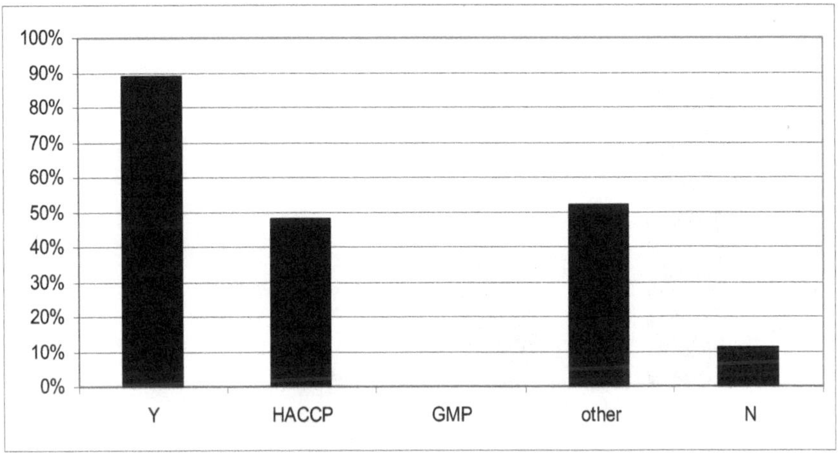

Figure 10. The existence of food safety monitoring programs

Question 5: "Before accepting raw materials do you carry out inspections for any of the following: chemical contamination, microbiological contamination or radioactivity?" reviled that no company checks the raw material for radiation, the most frequent inspections are for chemical and microbiological contamination (Figure 11).

At question 10: "How quickly do you think that you can respond to a food safety problem?" most of the companies said they can respond immediately (Figure 12). Although this result is a very good one for the risk management, is also very doubtful. There aren't any certain dates or situations that could verify how quickly these companies can intervene in a food safety problem.

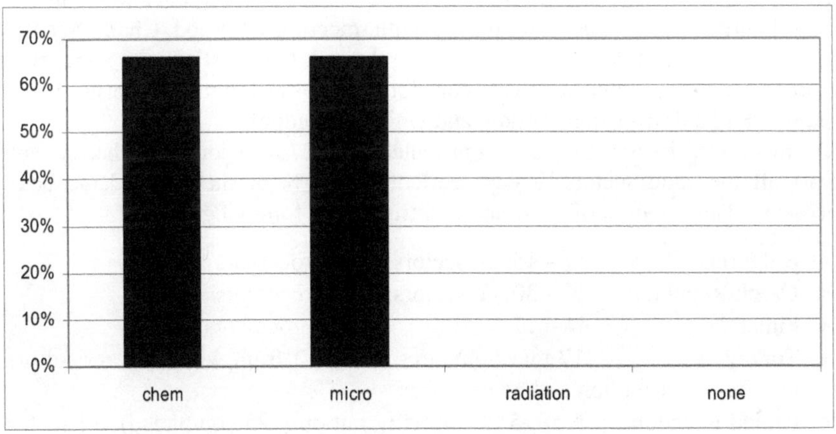

Figure 11. Inspections carried out by the companies before accepting the raw material

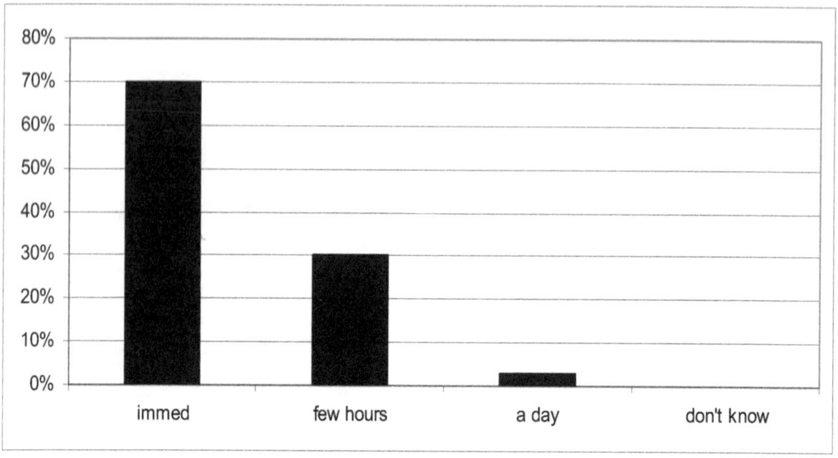

Figure 12. Necessary time to for the company to respond in a food safety problem

At question 11: "*Do you have crisis management system?*", 81% said that thy have one meanwhile 19% do not.

Questions 14 and 15 are about the presence of recall systems and if they had been tested or not. According to the answers received: 96% have a recall system, 74% even tested it and 4% don't have one yet.

7. Romanian Data in Comparison with Other Countries

Romania's contribution at the pilot study is quite recent. The data was gathered during February–May 2007. In the seventh meeting of Food Chain Security Project, in October 2007, all the dates obtain through the questionnaire so far were presented and the countries that contributed are Belgium, Bulgaria, Czech Republic, Finland, Romania, Turkey and United Kingdom.

As mentioned before, Romania contributed with 27 questionnaires that covered almost all the food sectors, except agriculture, 56% of the respondents being processors. The situation of the other countries is the following:

- Bulgaria: N = 15, all sectors, 80% processors
- Czech Republic: N = 30, all sectors, 30% processors
- Finland: N = 17, 100% processors
- Turkey: 17 questionnaires (N = 17) from which 15 agriculture, two from distribution and no producer
- United Kingdom: N = 25 (47 initially but only 25 completed), all sectors except agriculture, 100% processors

Regarding the products (dairy, meat, cereals, vegetables, fruits, beverages) the questionnaire completed by the Romanian companies included all categories except vegetables and no one specified at "other". The same was in Finland case, all categories of products were included except fruits, and other products were vinegar and confectionary. Bulgaria also included all products from the questionnaire meanwhile Turkey only dairy, cereals and vegetables. United Kingdom among the products already existing in the questionnaire, all except beverages, gathered data for bakery goods, snack foods nuts, pastry and so on. Czech Republic also added many new products like bakery goods, eggs, fish etc.

About the companies (questions 6–9) that participated, in Romania 66% were national companies and 34% multinational. For the other countries:

- Bulgaria: 73% national, 27% multinational
- Czech Republic: 67% national, 3% multinational
- Finland: not available
- Turkey: 65% national, 41% multinational
- United Kingdom: 78% national, 21% multinational

As it can be noticed, the Romanian companies' type is similar with the one from Czech Republic. Regarding the suppliers, everyone buys internationally: 71% Turkey, 4% UK (68% with both national and international suppliers), Finland also has international suppliers, as for Romania, 55% of the companies have national suppliers, 30% rather buy international and 15% have both types of suppliers.

About 72% (UK) and 82% (Czech Republic) is directly responsible with the food storage. In UK 6% have a contractor that is in charge with the storage and 19% share the responsibility. Concerning the distribution, 18% of the participant companies from Czech Republic and 47% from Turkey, UK and Bulgaria is directly responsible. Between 30% (UK) and 82% (Turkey) have a contractor and around 18% (Czech Republic) and 23% (UK) chosen both.

In Romania, 59% are personally responsible with the storage and 33% with distribution, 19% have a contractor in storages their products and 52% have one for distribution, 22% and 15% choused both answers when it comes to storage or distribution.

Hazard appreciation represents an important issue for food security. Question 1, 12 and 13 are looking for gathering data for this aspect. In Figure 13 are showed the most important risks for food stuffs in the participant companies' opinion. For each country is represented in the graphic only the sector with the most answers, in Romania's case, is processing.

As it can be noticed, the chemical and microbiological contamination is considered to be the greatest risks in all six countries (Figure 14):

- Chemicals: 12% (Czech Republic) – 80% (Bulgaria and UK)
- Bacteria/viral contamination: 6% (Turkey) – 76% (UK)
- Raw material: 13% (Czech Republic and Bulgaria) – 72% (UK)

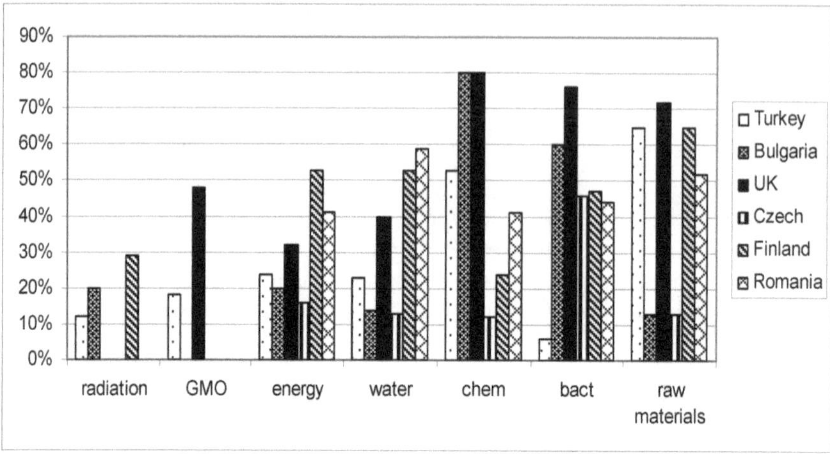

Figure 13. Hazards for food stuffs (question 1)

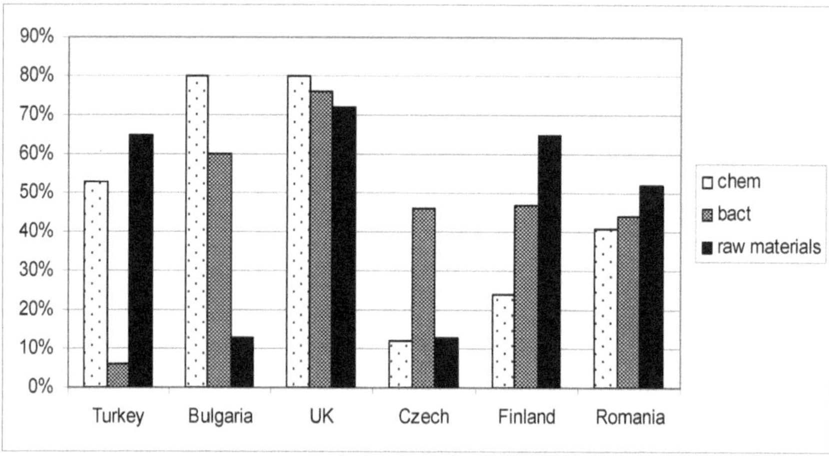

Figure 14. The most important risks for food security

The least concerns seem to be:

- Radiation: Turkey 12%, Bulgaria 20%, Finland 29%, the rest 0%
- Energy crisis: 16% (Czech Republic) – 53% (Finland)
- Water crisis: 13% (Czech Republic and Bulgaria) – 59% (Romania)
- GMO: UK (48%) and Turkey (18%), the rest 0% (Figure 15)

If raw material was considered one of the most important risks for food security the same happens at question 12, where is also considered to be the most vulnerable (38% in Czech Republic, 94% in Finland).

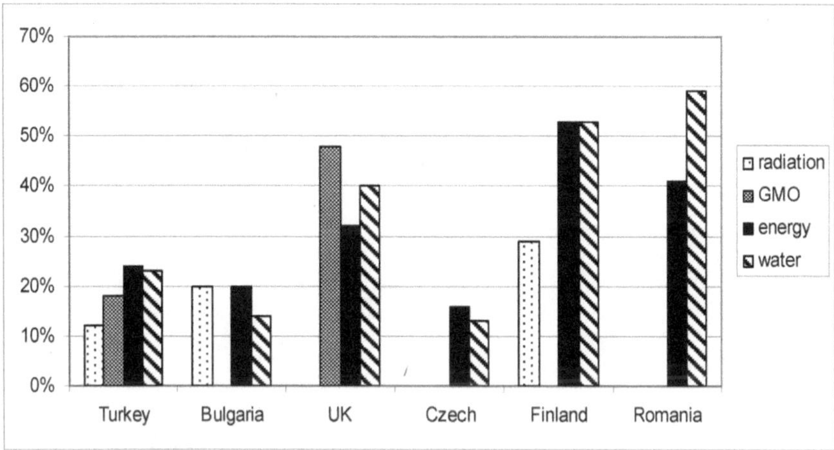

Figure 15. The less concerning risks for food security

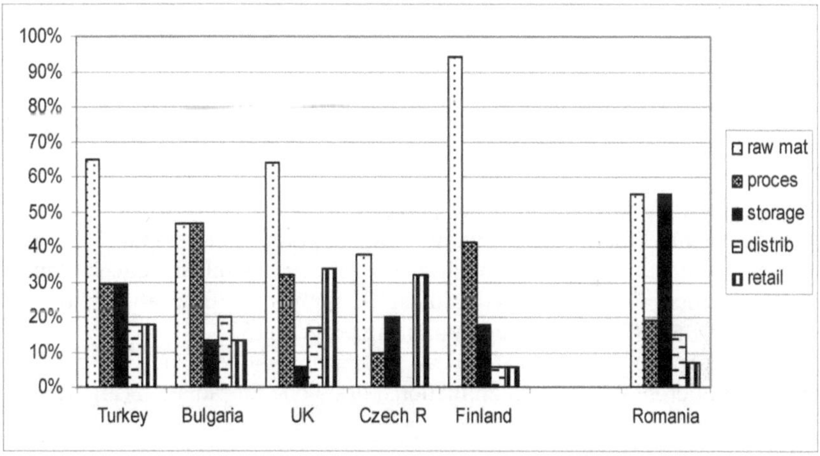

Figure 16. Vulnerable elements for the food security

Figure 16 show that beside Finland, Turkey and United Kingdom who considered raw material as being the most vulnerable, the other answers don't pass 55%.

Concerns about GMOs have a higher value only in UK and Turkey with 49% and 65% (Figure 17).

In risk management, "Food Chain Security" focuses also on the possible terrorist attacks that can affect food security. At the second question, if there are internal discussions about terrorist attacks, the situation turn showed that:

- Bulgaria Yes 27%, No 73%
- Czech Republic Yes 15%, No 85%
- Romania Yes 0%, No 100%
- Turkey Yes 53%, No 47%
- United Kingdom Yes 8%, No 92%

In Romania although the participant companies don't have any discussions about terrorist attacks around 70% verifies their employees from start. The same for Bulgaria, where all companies said they check the records of their employees. In UK only 36% of respondents do this kind of checking and 64% don't. In Turkey 94% check the employee's records and only 6% said no.

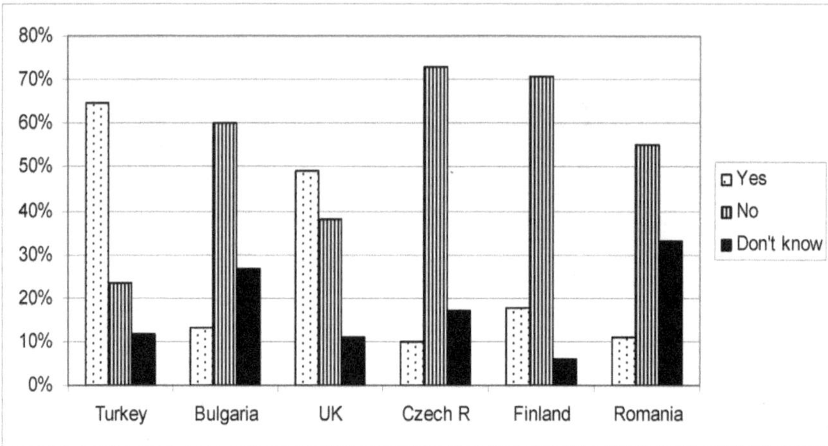

Figure 17. Do you think GMO is a safety concern

In Bulgaria, Turkey and UK the percentage is the same when it comes to the existence of a food security monitoring program. In Romania, 89% of the respondents said that they have such a program meanwhile in Czech Republic only 25%. Where it was specified, almost all companies, from all countries had HCCP, other GMP (21%) or other monitoring program (52%).

Before accepting the raw material all the participant companies inspect it for chemical or microbiological contamination, but no one for radiation (Figure 18).

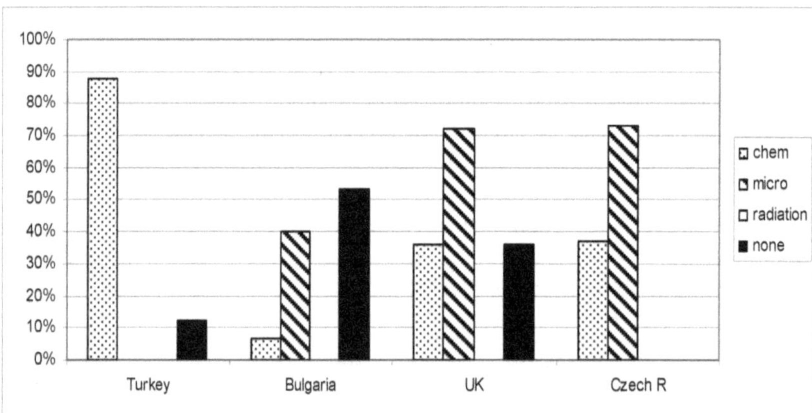

Figure 18. Controls done before accepting the raw materials (answers in Turkey have been only by Yes/No)

The majority of the respondents said that they have a crisis management system (question 11) as following:

- Turkey Yes 76%, No 24%
- Bulgaria Yes 60%, No 40%
- United Kingdom Yes 91%, No 9%
- Czech Republic Yes 66%, No 34%
- Romania Yes 81%, No 19%

Also, almost all the companies claimed to have recall system (questions 14 and 15) most of them already tested:

- Turkey Yes 88% No 12%; 86% tested/used
- Bulgaria Yes 53% No 47%; 13% tested/used
- United Kingdom Yes 98% No 2%; 94% tested/used
- Czech Republic Yes 50% No 50%; 50% tested/used
- Romania Yes 96% No 4%; 74% tested/used

Regarding the speed of intervention in case of a food safety problem (question 10) the majority said immediately or within a few hours:

- Turkey immediately/within a few hours 82%
- Bulgaria immediately/within a few hours 20%
- United Kingdom immediately/within a few hours 97%
- Czech Republic immediately/within a few hours 53%
- Romania immediately/within a few hours 97%

For the future, in Romania, we will move on to informing the public through articles in a local publication and a specialty one. Also, we have to identify things to look over and gathering new data where is needed, for confirming the existent ones. Knowing that traceability and communication where included in the questionnaire, they have to be determined from the competent authorities.

8. Concluding Remarks

During the last years, food security became a much bigger concern for the majority of the world's states; the authorities and many organizations are involving themselves in this matter. There fore, in 2003 a study had begun in Turkey, named "Food Chain Security" which is developing having NATO support and includes many states. Through this study, the analysis of the food security state is being searched and also future measures for the existing problems, or reducing the risk of new ones. For a statistic analysis of the food security, a questionnaire was proposed, that consists of 15 questions and which has been distributed to the participant countries. From the analysis of the dates that were gathered from Turkey, Bulgaria, Czech Republic, Great Britain, Belgium, Romania and Finland several conclusions can be shown:

1. The objectives of the study are to analyze the safety of food stuffs, and to search response measures that could be taken to reduce the risk and to mitigate the consequences of these threats for human health and well-being.
2. In a human ecosystem, alimentation has a major importance. Through metabolic processes it influences the individual who involuntary influences the environment.
3. Agriculture lays at the base of ensuring food security and it can be found among the majors consumers of ecosystems and their services, having also the greatest impact upon them (soil contamination with agrochemicals).
4. The problem of ensuring enough food for the population, in other words ensuring food security, must become a priority for the states that have as a long time objective, sustainable development.
5. Food security is part of a chapter in "The national strategy of Romania for a sustainable development" (21 Agenda), which means that the Romanian authorities started to take in count its importance for a sustainable development.
6. The interviewed companies from the south-east region of Romania, that completed the questionnaire, covered all the sectors of the food chain excepting agriculture and also many products. The majority were processors (56%) also I countries like UK (100%) and Finland (100%) but not necessarily, in Turkey for example from 17 questionnaires, 15 were from agriculture and none from processing.
7. The products covered with the questionnaire were of many types. None of the Romanian companies that completed the questionnaire, specified at "other products" although there were many answers here (30%). Even so, all types of products existing in the questionnaire were included, except vegetables. The rest of the countries included, among the existing types of products.
8. All the respondents have international supplier among the national ones, in Romania (30%) and in the rest of the countries: Turkey 71%, UK 4%.
9. Many companies share the storage and distribution responsibility with a contractor, or they don't get involved at all. When it comes to storage only 33% of the Romanian companies chose to be personally responsible with it, meanwhile in UK or Czech Republic this category was represented by 72% and 82% of the companies, which means that the our companies are not preoccupied or they are but not enough, with the risks that can appear in the food chain and endanger food security.
10. Chemical and microbiological contamination and also raw material contamination are considered to be the most important risks for food security, by all countries in different percentages, meanwhile, radiations, GMO, energy and water crisis are being less considered as possible risks. This situation shows that is possible that those were incorrect evaluate.
11. Only the companies from Romania considered water crisis as a risk in a higher percentage (59%) and Finland for energy crisis (53%).

12. Raw material remains in everyone's opinion the most vulnerable element for food security (55% Romania – 94% Finland).
13. No one inspects the raw material for radiation, before accepting it. The most often inspection are microbiological (66% in Romania, over 70% in UK and Czech Republic) and chemicals (Bulgaria less than 10%, in Romania the same percentage like microbiological inspections, some companies carrying out both types).
14. Food security monitoring programs exists in all the countries: 100% in Turkey, Bulgaria or United Kingdom, 89% in Romania.
15. Crisis management systems exist among all the participant companies: 81% in Romania, 91% in UK.
16. Most of the companies have recall systems for their products in case of necessity, used or just tested (UK 98% from which 94% tested or used, Romania 96% from which 76% tested/used) and they affirmed they could intervene immediately or within a few hours (Romania and UK – 97%).
17. There aren't too many concerns regarding terrorist attacks. In Romania, as it come out from the questionnaire, no company has this kind of discussions meanwhile in Turkey, 53% said they do.
18. A value of 70% of the Romanian respondents claimed that the check their employees records, from the start, this being a good thing considering that the record's check can diminish the risks of terrorist attacks. In other countries, these checks are done by all the companies (100% in Bulgaria) or a quite small percentage of them (36% in UK).

Considering the fact that the numbers of the questionnaires from our country is relatively small, there are some doubts about the results extend outside the group. There fore, they need to be confirmed by supplementary information or with new data.

Despite this situation, the existing results showed two aspects of great importance for food security:

- Food security management systems are implemented in the existing group, in some countries even 100% and in Romania, 89% of the participant companies own this type of systems.
- Crisis management systems and recall systems in case of emergency exist, are being used and can intervene very quickly.

For Romania to talk about a sustainable development, food security must become a priority and in the shortest time, a real thing. Romania can contribute to different studies that take place today around the world in search of improving the state of food chain security based on analyses, measures of preventing and indication given by them. Ensuring food security can influence human well-being, the economy, human health but it can also mean the correct utilization of the natural capital resources, all this being objectives of the sustainable development as it can be observed from 21 Agenda.

Acknowledgments

We are thankful to our colleagues from different countries acting as the members of the Food Chain Security Pilot Study, who kindly shared information with us about their research and results of their work, all along these years. Based on the work of Momtchil Sidjimov (Bulgaria), Ales Komar and Pavlina Vasicka (Czech Republic), Hannu Sivonen and Kyösti Orre (Finland), Hami Alpas, Beyazit Cirakoglu and Faruk Bozoglu (Turkey), Madeleine Smith (UK), we have been able to write chapter 7, Romanian Data in Comparison with Other Countries. We are grateful for their kindness.

We also thank Professor Stoica-Preda Godeanu for his helpful and suggestions during our research in Romania.

References

World-Watch Institute, 2008, *State of the World* - Meat and Seafood: The Global Diet's Most Costly Ingredients, 63–71.
Romanian National Institute of Statistics, 2005, Anuarul statistic al României.
Romanian National Institute of Statistics, 2006, Anuarul statistic al României.
Romanian National Institute of Statistics, 2007, Anuarul statistic al României.
Guvernul României, 2008, Strategia Naţională a României pentru Dezvoltare Durabilă, Editura NOVA, Bucureşti, 15–17, 41–43.
Romanian National Institute of Statistics, 2008, Anuarul statistic al României.

New System of Food Control in Russia

Irina V. Ermakova

Institution of Russian Academy of Sciences, Institute of Higher Nervous Activity and Neurophysiology of RAS, Butlerov str., 5a, Moscow 117485, Russian Federation

Abstract

Food safety is quite important for human health in all countries. Humanity has the uniform space and must take care about all parts of it. Pollution of one region leads to the same state of others regions through water, land, air, living organisms. The poor-quality or poison ed food products can pollute the territory and influence negatively on the environment. The food security is important, especially, in connection with the possibility of terrorist attacks. The Federal Service of control in sphere of protection of the rights of consumers was formed in Russia in 2004. This Service carries out the activity directly and through the territorial organizations in interaction with administrative structures in food control and inspection, namely in sanitary-epidemiological service, veterinary service, grain service, inspection of trade connections and standardization and certification. The control is carried out on the basis of laws. The law № 29-FZ concerns the quality and safety of foodstuff and how to control it. The law 134-FZ attracted the protection of the rights of legal persons and individual businessmen. The modification of the law №234-FZ about protection of the rights of consumers is connected with new food – GMOs. Great attention is paid to the safety of new food. Private companies also perform analysis of heavy metals, mycotoxins, radiation and the presence of bacteria, virus or genetically modified organisms. We would like to thank the group Pilot Study "Food chain security" for very important work concerning food safety in different countries. They help us to understand internal problems in Russia and to create the cooperation with other countries. All these steps are very important for the protection population from toxic food.

1. Why the Food Safety Is Important?

Food safety is a scientific discipline describing the handling, preparation, and storage of food in ways that prevent food borne illness. This includes a number of routines that should be followed to avoid potentially severe health hazards. It was shown that poor-quality or poisoning food can lead to the premature ageing and

H. Alpas and B. Cirakoglu (eds.), *Food Chain Security*,
DOI 10.1007/978-90-481-9558-9_6, © Springer Science + Business Media B.V. 2010

mental diseases as epilepsy, schizophrenia, Alzheimer and Parkinson diseases, and others. One of the reasons of these kinds of diseases is the disbalance between inhibitory and excitatory neurotransmitters caused by small amount of neurotoxic substances in food.

Food can transmit disease from person to person as well as serve as a growth medium for bacteria that can cause food poisoning. Debates on genetic food safety include such issues as impact of genetically modified food on health of further generations and genetic pollution of environment, which can destroy natural biological diversity. In developed countries there are intricate standards for food preparation, whereas in lesser developed countries the main issue is simply the availability of adequate safe water, which is usually a critical item.

2. New System of Food Control

2.1. Federal Service (from www.rospotrebnadzor.ru)

The control for the food security was very strong in Soviet Union. However this system was destroyed during the period of Perestroika. New system of food control taking into account democratic rules was created in Russia last years. The Federal Service of control in sphere of protection of the rights of consumers was formed in 2004. This Service carries out the activity directly and through the territorial organizations in interaction with administrative structures. Eighty-nine territorial administrations and 90 centers of hygiene and epidemiology in subjects of the Russian Federation were formed. Experts of Federal Service carry out the sanitary-quarantine control in 285 check points, including 102 – on motor transport, 67 – at the airports, 64 – sea, 13 river, 39 – at boundary railway stations. Federal Service has 28 scientific Institutes, 14 anti plague stations, 14 disinfection organizations. More than 110,000 specialists worked there. This Service has received some functions from the Ministry of health, the Ministry of economic development and the trade, and the Ministry of an antimonopoly policy.

2.2. Legal Basis (www.rospotrebnadzor.ru)

Legal bases of activity of Federal Agency are established by federal laws: "About sanitary-and-epidemiologic well-being of the population", "About protection of the rights of consumers" and other standard legal certificates of the Russian Federation. The control for food safety is carried out on the basis of 43 laws.

Important information of food safety was reflected in the following laws:

The law № 29-FZ "About quality and safety of foodstuff", that was created 02.01.2000 and changed on December 30, 2001, on January 10, on June 30, 2003, on August 22, 2004, on May 9, 2005, on December 31, 2005, on March 31, 2006 (see below the details).

The law 134-FZ about the protection of the rights of legal persons and individual businessmen at carrying out of the state control regulates the interaction between the individuals and state carried out in 2001 with changes from October 30, 2002, on January 10, on October 1, 2003, on August 22, 2004, on May 9, on July 2, on December 31, 2005. Main principles of protection of the rights of legal bodies and individual businessmen are: a presumption of conscientiousness of the legal person or the individual businessman; observance of the international contracts of the Russian Federation; the openness and availability to legal bodies and individual businessmen of the standard legal certificates establishing obligatory requirements, which performance is checked at carrying out of the state control (supervision); an establishment of obligatory requirements federal laws and the standard legal certificates accepted according to them; carrying out of actions for the control the authorized officials of bodies of the state control (supervision); conformity of a subject of carried out action for the control of the competence of body of the state control (supervision); periodicity and efficiency of carrying out of action for the control, providing full and its as much as possible fast carrying out during a target date; the account of actions for the control, spent by bodies of the state control (supervision); possibility of the appeal of actions (inactivity) of officials of bodies of the state control (supervision) breaking an order of carrying out of actions under the control, established by the present federal law, other federal laws and the standard legal certificates accepted according to them; a recognition in an order established by the federal legislation, invalid (in full or in part) the standard legal certificates establishing obligatory requirements which observance is subject to check if they do not correspond to federal laws; elimination in full bodies of the state control (supervision) of the admitted infringements in case of a recognition court of the complaint of the legal person or the individual businessman of the proved; responsibility of bodies of the state control (supervision) and their officials at carrying out of the state control (supervision) for infringement of the legislation of the Russian Federation; inadmissibility of collection by bodies of the state control (supervision) of a payment from legal bodies and individual businessmen for carrying out of actions for the control, except for cases of the reimbursement of bodies of the state control (supervision) on realization of researches (tests) and examinations in which result infringements of obligatory requirements are revealed; Inadmissibility of direct reception by bodies of the state control (supervision) of deductions from the sums collected from legal bodies and (or) individual businessmen as a result of carrying out of actions under the control.

Modification of the Law of the Russian Federation "About protection of the rights of consumers" in connection with new food – genetically modified food, 12.12.2007. Thereupon the Federal Agency of supervision in sphere of protection of the rights of consumers and well-being of the person with a view of formation uniform legal experts considers necessary to pay attention to below-mentioned separate positions of the specified act. The addition imperatively fixing the basis for instructions without fail concerning a foodstuff of the information on presence in them of components, the gene-engineering-modified organisms received with application is made to the Law (GMO), – "in case the maintenance of the specified organisms in such component makes more the than nine tenth percent". Thus, taking into account objective necessity of definition of an order of corresponding marking of the foodstuff received from GMO as forms of realization of the right of the consumer on timely reception necessary and a trustworthy information about structure of the foodstuff, providing possibility of their correct choice, the Law has been harmonized with requirements of the European Union on labeling the foodstuff received from GMO, established by the Instruction of the European Parliament and Council from 22.09.2003 № 1829/2003 about genetically modified food and forages which since April 2004 has entered 0.9% threshold level for marks of the foodstuff received from GMO. Thus it is necessary to notice that the maintenance in foodstuff of 0.9% and less the components received with application GMO, is casual or technically ineradicable impurity and the foodstuff containing specified quantity of such components, do not concern a category of the foodstuff containing components, received with application GMO.

2.3. About Quality and Safety of Foodstuff (Law № 29-FZ)

Inspection of raw materials on radioactivity, chemical and biologic agents is carried out before accepting of this material. Great attention is paid to such risks as concentration of pesticides, heavy metals and other chemicals, presence of bacteria/virus, micotoxins, level of radiation, nitrites – nitrates, GM-components in products.

2.3.1. Ability to the Turn and Advancement of Foodstuff, Materials and Products

1. In a turn there can be foodstuff, materials and the products which corresponding to requirements of standard documents and have passed the state registration in an order, established by the present Federal law.

2. There can not be in a turn foodstuff, materials and products, which

(a) Do not correspond to requirements of standard documents.

(b) Have no certificates of quality and safety of foodstuff, materials and products, documents of the manufacturer, the supplier of foodstuff, materials and the products confirming their origin in which relation there is no information on the state registration and acknowledgement of conformity to requirements of standard documents (foodstuff, materials and the products which are subject to the state registration and obligatory acknowledgement of conformity).

(c) Do not correspond to the presented information and in which relation there are well-founded suspicions about their falsification; have no validity target dates (for foodstuff, materials and products in which relation the establishment of working lives is obligatory) or which working lives have expired.

(d) Have no the marks containing data, provided the law or in state standard or in which relation there is no such information. Such foodstuff, materials and products admit poor-quality and dangerous and realizations are not subject, are utilized or destroyed.

2.3.2. Quality and Safety of Foodstuff

Maintenance of quality and safety of foodstuff, materials and products quality and safety of foodstuff, materials and products are provided with means: applications of measures of state regulation in the field of maintenance of quality and safety of foodstuff, materials and products; carrying out by citizens, including individual businessmen, and the legal bodies who are carrying out activity on manufacturing and a turn of foodstuff, materials and products, organizational, agrochemical, veterinary, technological, technical, sanitary-antiepidemic and phytosanitory actions for performance of requirements of standard documents to foodstuff, materials and products, conditions of their manufacturing, storage, transportations and realization; Industrial inspection carrying out behind quality and safety of foodstuff, materials and products, conditions of their manufacturing, storage, transportations and realizations, introduction of control systems by quality of foodstuff, materials and products (further – quality systems); applications of measures on suppression of infringements of the present Federal Law, including requirements of standard documents, and also measures civil-law, administrative and the criminal liability to the persons guilty of fulfillment of specified infringements.

2.3.3. The Information on Quality and Safety of Foodstuff, Materials and Products

1. Individual businessmen and the legal bodies who are carrying out activity on manufacturing and a turn of foodstuff, materials and products, to rendering of services in sphere of retail trade in foodstuff, materials and products and public catering sphere, are obliged to give to buyers or consumers, and also bodies of the state supervision and the control full both a trustworthy information about quality

and safety of foodstuff, materials and products, observance of requirements of standard documents at manufacturing and a turn of foodstuff, materials and products and rendering of such services.

2. Federal enforcement authority on the state supervision in the field of standardization and certification, federal enforcement authority in the field of the state sanitary-and-epidemiologic supervision, federal enforcement authority in the field of the state veterinary supervision public authorities provide, citizens (including individual businessmen) and legal bodies with the information on quality and safety of foodstuff, materials and products, about observance of requirements of standard documents at manufacturing and a turn of foodstuff, materials and products, rendering of services in sphere of retail trade in foodstuff, materials and products and public catering sphere, about the state registration of foodstuff, materials and products, about acknowledgement of their conformity to requirements of standard documents, And also about standard documents and measures on prevention of realization of poor-quality and dangerous foodstuff, materials and products.

3. Control for Safety of Foodstuff and Health of the Population

1. Powers of the Russian Federation in the field of maintenance of quality and safety of foodstuff concern: working out and carrying out in the Russian Federation a uniform state policy; acceptance of federal laws and other standard legal certificates of the Russian Federation; working out and realization of federal target and scientific and technical programs of maintenance of quality and safety of foodstuff, materials and products; the state rationing in the field of maintenance of quality and safety of foodstuff, materials and products; the organization and realization of the state registration of foodstuff, materials and products; The organization and carrying out of obligatory certification of separate kinds of foodstuff, materials and products, and also the services rendered in sphere of retail trade by foodstuff and sphere of public catering, quality systems; the organization and carrying out of the state supervision and the control; realization of the international cooperation of the Russian Federation; realization of others provided by the legislation of the Russian Federation of powers.

2. Public authorities of subjects of the Russian Federation have the right to participate in realization of powers of the Russian Federation in the field of maintenance of quality and safety of foodstuff means: acceptances according to federal laws of laws and other standard legal certificates of subjects of the Russian Federation; workings out, statements and realizations of regional programs of maintenance of quality and safety of foodstuff; Realization together with the authorized federal enforcement authorities of the control and supervision of quality and safety of foodstuff.

3. With a view of definition of priority directions of a state policy in the field of maintenance of quality and safety of foodstuff, public health care, and also with a view of working out of measures on receipt prevention on the consumer market of poor-quality and dangerous foodstuff, materials and products bodies of the state supervision and the control together with enforcement authorities of subjects of the Russian Federation will organize and carry out monitoring of quality and safety of foodstuff, population health. Monitoring of quality and safety of foodstuff, population health is spent according to the position confirmed by the Government of the Russian Federation.

3.1. New Food: GMOs

According to the legislation of the Russian Federation (Federal laws from 05.07.1996 № 86-FZ "About state regulation in the field of gene-engineering activity", from 02.01.2000 № 29-FZ "About quality and safety of foodstuff" and from 30.03.1999 № 52-FZ "About sanitary-and-epidemiologic well-being of the population") food production from GMO concerns a category of "new food" and is subject to an obligatory estimation on safety and to the subsequent monitoring behind a turn.

The term genetically modified organisms (GMOs) refers to plants, microbes and animals with genes transferred from other species in order to produce certain novel characteristics (for example resistance to pests, or herbicides), and are produced by recombinant DNA technology. There are two standard methods, which are generally used to introduce new DNA (genes) into a plant cell, which is going to be modified: (1) the particle acceleration, or "shot-gun" technique, and (2) infecting the cells with a modified pathogen, with the help of *Agrobacter tumefaciensis*. Neither methods are perfect nor do not guarantee that the rest of the plant genom remains unchanged. Therefore the safety of the GM-crops created with the help of these methods cannot be garanteed neither for human and animal health, nor forth the environment (Ho and Tappeser 1997; Kuznetcov et al. 2004; Wilson et al. 2004; Ermakova 2005). Four main sources of the hazards of GMO are accepted by scientists worldwide: (1) those due to the new genes, and gene products introduced; (2) unintended effects inherent to the technology; (3) interactions between foreign genes and host genes; and (4) those arising from the spread of the introduced genes by ordinary cross-pollination as well as by horizontal gene transfer (World Scientists' Statement 2000).

GM-crops contain material, which under natural conditions is not present in them in nature, and they form a part of our daily diet. To understand what effect they can have on us and on our animals it is vitally important to study the influence of these GM-plants in different organisms for several generations. At the present, these studies are lacking from the scientific literature. Also, several detrimental effects of GM-crops had been showed on the metabolism of animals.

The hazard of genetically modified organisms (GMO) was shown for animals and the environment in many investigations (Traavik 1995; Ho and Tappeser 1997; Pusztai 1999, 2001; Kuznetcov et al. 2004 and others). Earlier it was shown that consumption of GM-food by animals led to the negative changes in their organisms. Experiments, conducted by Arpad Pusztai showed that potatoes modified by the insertion of the gene of the snowdrop lectin (insecticidal proteins), stunted the growth of rats, significantly affected some of their vital organs, including the kidneys, thymus, gastrocnemius muscle and others (1998) and damaged their intestines and their immune system (Ewen and Pusztai 1999). Similar effect of GM-potatoes on rats was obtained at Institute of Nutrition in Russia (Medical-biological Report 1998). Significant modifications in the cells of liver, exocrine pancreas and testis of mice, fed by diet containing Roundup Ready GM-soybean were shown by Malatesta with coauthors (Malatesta et al. 2002, 2003; Vecchio et al. 2004). In another investigation influence of GM-pear on mice was found (Prescott et al. 2005). French and Austrian investigators showed the negative effect of GM corn (Seralini et al. 2007; Velimirov et al. 2008).

English scientist Arpad Pusztai in his article (2001) asked: “How can the public make informed decisions about GM foods when there is so little information about its safety?”

It is put forward in the risk assessment documents that the GM-components of transformed plants are completely destroyed in the digestive tract of humans and animals, together with the other genetic material found in them. However foreign DNA plasmids are steadier against the digestion, than it was originally believed. Plasmid DNA and GM-DNA were found in microorganisms of the intestine and in saliva (Mercer et al. 1998; Coghlan 2002). Experimental researches in mice showed that ingested foreign DNA can persist in fragmented form in the gastrointestinal tract, penetrate the intestinal wall, and reach the nuclei of leukocytes, spleen and liver cells (Schubbert et al. 1994). In another research of Shubbert et al. (1998) the plasmid containing the gene for the green fluorescent protein (pEGFP-C1) or bacteriofaphage M13 DNA were fed to pregnant mice. Foreign DNA, orally ingested by pregnant mice, was discovered in blood (leukocytes), spleen, liver, heart, brain, testes and other organs of foetuses and newborn animals. The authors considered that maternally ingested foreign DNA could be potential mutagens for the developing fetus. High mortality and infertility of offspring from the parents fed by GM-soy were found in the investigations of Russian scientists (Ermakova 2006, 2007; Malygin 2008 and others). Obtained data allow to presume that the negative effect of GM-soy on the newborn pups could be mediated by three possible factors. Firstly, it can be the result of transformation, and insertion of foreign DNA (transgenes, plasmids and etc.), which could penetrate into the sexual/stem cells, or/and into cells of the fetus, as it was observed by Schubbert and colleagues (1998). Secondly, the negative effect of GM-soya could be connected with the highly mutagenic nature of the GM transformation process. Also, the instability of gene constructs was described for GM-soya (Windels et al. 2001). Thirdly, the adverse effect of GM-soya could be mediated by the accumulation of roundup residues in GM-soya. However, no mortality was observed with female rats, nor with the young pups survived, although they also began to eat the GM-soya, it was supposed that the effect could be mediated by two first factors. Confirmation of this assumption could be also the fact of weak negative effect of protein-isolate GM-soya.

At the present time one of the main global goals is the protection of an environment from the uncontrolled distribution and the contamination of imperfect genetically modified organisms, which can cause the human diseases, the decrease of biovariety, and the destruction of nature. It was shown also that transgenic crops are not compatible with agroecologically based systems of production: the myth of coexistence (Altieri 2005).

It is admitted to industrial production more than 100 lines of genetically modified plants in the world. In the Russian Federation there is a necessary is standard-methodical base, including the methods of laboratory researches necessary for the organization and carrying out of effective supervision of foodstuff, containing the components received with application GMO. According to the legislation of the Russian Federation (Federal laws from 05.07.1996 № 86-FZ "About state regulation in the field of gene-engineering activity", from 02.01.2000 № 29-FZ "About quality and safety of foodstuff" and from 30.03.1999 № 52-FZ "About sanitary-and-epidemiologic well-being of the population") food production from GMO concerns a category of "new food" and is subject to an obligatory estimation on safety and to the subsequent monitoring behind a turn. The foodstuff received from GMO, passed a medical and biologic estimation and not different on the studied properties from the analogues received by traditional methods, is safe for health of the person, are resolved for realization to the population and use in the food-processing industry without restrictions. In the world there are different lines of thought to marks of the foodstuff received from GMO, however it is not connected with a safety issue of the foodstuff received from GMO, and pursues the aims of informing of the population about use of concrete technology of data acquisition of products. According to the letter of Rospotrebnadzora from 24.01.2006 № 0100/446-06-32 maintenance in foodstuff of 0.9% and less the components received with application GMO, are casual or technically ineradicable impurity and the foodstuff containing specified quantity of components GMO, do not concern a category of the foodstuff containing components, received with application GMO. Several lines of GM cultures have passed a full cycle of all necessary researches and are resolved for use in the Russian Federation: 14 lines of food production, a phytogenesis received with application of transgene technologies: six lines of corn, three lines of a soya, three lines of a potato, one line of a sugar beet, one line of rice; five kinds of genetically modified microorganisms. Researches of foodstuff on presence of GMOs are spent according to state standards of the Russian Federation of GOST P 52173-2003 "Raw materials and products food. A method of identification of genetically modified sources (GMS) a phytogenesis" and GOST P 52174-2003 "Raw materials and products food. A method of identification of genetically modified sources (GMS) a phytogenesis with application of a biological microchip", and also methodical instructions 4.2.1913-04 "Methods of quantitative definition of genetically modified sources (GMS) a phytogenesis in a foodstuff", confirmed by the Main state health officer of the Russian Federation. In 2004 the main state health officer of the Russian Federation confirms different methodical instructions: Methodical instructions 4.2. 1917-04 "the Order and the organization of the control over the food production received from/or with use of raw materials of a phytogenesis, having genetically modified analogues"; – Methodical instructions 2.3.2.1935-04 "Order and the organization of the control over the food production received from/or with use of genetically modified microorganisms and microorganisms, having genetically modified analogues". – Methodical instructions 4.2. 1902-04 "Methods of definition of concrete lines GMO in foodstuff". – Methodical instruction 2.3.2. 1830-04 "the Microbiological and molecular-genetic estimation of the food production received with use of genetically modified microorganisms". The decision of the main state health officer of the Russian Federation from 31.12.2004 № 13 "About strengthening of supervision of the foodstuff received from GMOs" defines the head centers by quantitative definition GMO in a foodstuff in federal districts of the Russian Federation which functions include completion by the necessary equipment, training of experts to methods and carrying out of researches under quantitative maintenance GMO in foodstuff in territory of corresponding federal district of the Russian Federation. In 2005 territorial administrations of Federal Agency of supervision in sphere of protection of the rights of consumers and well-being of the person on subjects of the Russian Federation on presence of the components received with application GMO, investigate 18872 tests of food raw materials and foodstuff (2004 – 12956, 2003 – 4300).

4. Conclusion

It is quite important to pay attention on the food safety. Especially the food security is important in connection with the possibility of terrorist attacks.

The management of control for food safety is rather strict in Russia at the present time. The interrogation of 270 companies by using the questionnaire made by Committee "Food chain security" (carried out twice: in 2007 and 2008yy) revealed the high activity of the companies to the food safety. The companies paid main attention to the presence of chemical substances, bacteria/virus and GMOs in their products. Most companies have a food safety monitoring program and a recall system for the products. However the insufficient control in some cases has resulted in increase of different diseases as oncology of internal organs, allergy, toxic reaction, etc. found in population last years.

Special problem is connected with the wide distribution of GM-products in Russia. However first steps were made already. Local laboratories were created in different cities and towns in Russia for GMO control last years. Sixteen special laboratories were organized for GMO control by using polymerase chain reaction (PCR) only in Moscow. GMO free zones were declared in different regions. Obligatory marks of presence of GMO more than 0.9% were accepted. Label "GMO free" was recommended to companies also.

We can conclude that the development of democratic system in Russia allowed creating new rules and enforcing the control for food safety in our country both by state and different organizations and persons.

References

Altieri M.A. The myth of coexistence: why transgenic crops are not compatible with agroecologically based systems of production. Bulletin of Science, Technology and Society, 25, 4, 2005, pp.361–371.

Brake D.G. and Evenson D.P. A generational study of glyphosate-tolerant soybeans on mouse fetal, postnatal, pubertal and adult testicular development. Food Chemistry and Toxicology 42, 2004, pp.29–36.

Coghlan A. GM crop DNA found in human gut bugs. NewScientist. 2002.

Ermakova I. Influence of genetically modified soya on the birth-weight and survival of rat pups// Proceedings "Epigenetics, Transgenic Plants and Risk Assessment", 2006. pp.41–48.

Ermakova I.V. GM soybeans – revisiting a controversial format. Nature Biotechnology, 25(12), 2007, pp.1351–1354.

Ho M.W. and Tappeser B. Potential contributions of horizontal gene transfer to the transboundary movement of living modified organisms resulting from modern biotechnology. In Transboundary Movement of Living Modified Organisms Resulting from Modern Biotechnology: Issues and Opportunities for Policy-Makers (K.J. Mulongoy, ed.) International Academy of the Environment, Switzerland, 1997, pp.171–193.

Kuznetcov V.V., Kulikov A.M., Mitrohin I.A. and Cidendambaev V.D. "Genetically modified organisms and biological safety" Ecos, 2004, pp.3–64.

Mercer, D.K., Scott, K.P., Bruce-Johnson, W.A., Glover, L.A. and Flint, H.J. Fate of free DNA and transformation of oral bacterium Streptococcus gordonii DL1 plasmid DNA in human saliva. Applied and Environmental Microbiology, 65, 1999, pp.6–10.

Malatesta M., Biggiogera M., Manuali E., Rocchi M.B.L., Baldelli B., Gazzanelli G: Fine structural analyses of pancreatic acinar cell nuclei from mice fed on GM soybean. European Journal of Histochemistry, 47, 2003, pp.385–388.

Malatesta M., Caporalony C., Gavaudan S., Rocchi M.B.L., Tiberi C., Gazzanelli G. Ultrastructural, morphometrical and immunocytochemical analysis of hepatocyte nuclei from mice fed on genetically modified soybean. Cell Structure and Function, 27, 2002, pp.173–180.

Official website: www.rospotrebnadzor.ru

Pusztai A. Genetically Modified Foods: Are They a Risk to Human/Animal Health. Biotechnology: genetically modified organisms. 2001.

Schubbert R., Lettmann C., Doerfler W. Ingested foreign (phage M13) DNA survives transiently in the gastrointestinal tract and enters the blood stream of mice. Molecules, Genes and Genetics, 242, 1994, pp.495–504.

Schubbert R., Hohlweg U., Renz D., Doerfler W. On the fate of orally ingested foreign DNA in mice: chromosomal association and placental transmission in the fetus. Molecules, Genes and Genetics, 259, 1998, pp.569–576.

Seralini G.E., Cellier D., Vendomois JS. New analysis of a rat feeding study with a genetically modified maize reveals signs of hepatorenal toxicity. Archives of Environmental Contamination and Toxicology, 52(4), 2007, pp.596–602.

Traavik, T. Too Early May Be Too Late. Ecological Risks Associated with the Use of Naked DNA as a Biological Tool for Research, Production and Therapy (Norwegian), Report for the Directorate for Nature Research Tungasletta 2, 7005 Trondheim. English translation, 1999.

Vecchio L., Cisterna B., Malatesta M., Martin T.E., Biggiogera B. Ultrastructural analysis of testes from mice fed on genetically modified soybean. European Journal of Histochemistry, 48, 2003, pp.449–453.

Velimirov A., Binter C., Zentek J. Biological effects of transgenic maize NK603xMON810 fed in long term reproduction studies in mice. Report, Forschungsberichte der Sektion IV, Band 3. Institut für Ernährung, and Forschungsinttitut für biologischen Landbau, Vienna, Austria, November 2008.

Windels P., Taverniers I., Depicker A., Van Bockstaele E., De Loose M. Characterisation of the Roundup Ready soybean insert. European Food Research and Technology, 231, 2001, pp.107–112.

World Scientists Statement. Supplementary Information of the Hazards of Genetic Engineering Biotechnology. Third World Network. 2000.

(World Scientists' Statement 2000): Supplementary Information on the Hazards of Genetic Engineering Biotechnology

Website of The Federal Service of control in sphere of protection of the rights of consumers http://www.rospotrebnadzor.ru

Food Chain Defense in the United States

LeeAnne Jackson*

*Contact Person

U.S. Food and Agriculture Sector

Health Science Policy Advisor, Food and Drug Administration, Center for Food Safety and Applied Nutrition, Office of Food Defense, Communication, and Emergency Response, Food Defense Oversight Team, 5100 Paint Branch Parkway HFS-007, College Park, MD 20740, USA

Abstract

Protecting the United States food and agricultural critical infrastructure and key resources is an important responsibility shared by Federal, State, local, tribal, and territorial governments and private industry partners. Interference with the food or agricultural infrastructure could have a devastating impact on the Nation's public health and economy. Adequate protection and resiliency of infrastructure in the Food and Agriculture Sector requires a number of integrated processes and procedures undertaken by all sector partners. A number of tools, guidance documents, educational materials, and regulations are available to aid in protecting the United States Food and Agriculture Sector.

1. Introduction

Protecting the critical infrastructure and key resources (CIKR) of the United States is essential to the Nation's security, economic vitality, and way of life. CIKR includes assets, systems, and networks that provide vital services to the Nation. Terrorist or other manmade attacks on CIKR and natural disasters, also known as the "all hazards" approach, could significantly disrupt the functioning of government and business alike and produce cascading effects far beyond the affected CIKR sector and physical location of the incident.

The protection of the Nation's CIKR, therefore, is an essential part of the homeland security mission of making America safer, more secure, and more resilient from terrorist attacks and other natural and manmade hazards. Protection includes actions to guard or shield CIKR assets, systems, networks, or their interconnecting links from exposure, injury, destruction, incapacitation, or exploitation. This also includes actions to deter, mitigate, or neutralize the consequence, vulnerability, or threat associated with a terrorist attack or other incident.

H. Alpas and B. Cirakoglu (eds.), *Food Chain Security*,

DOI 10.1007/978-90-481-9558-9_7, © Springer Science + Business Media B.V. 2010

2. Food and Agriculture Sector Profile

The United States Food and Agriculture Sector is composed of complex production, processing, and delivery systems and has the capacity to feed people and animals beyond the boundaries of the United States. These food and agriculture systems are almost entirely under private ownership, operate in highly competitive global markets, strive to operate in harmony with the environment, and provide economic opportunities and an improved quality of life for United States citizens and others worldwide. The sector accounts for roughly one-fifth of the United States economic activity. One-fifth of the agricultural production is exported, generating $115.5 billion in 2008, creating a positive trade balance of roughly $35 billion, and thereby fueling the US economy.

The United States has approximately 44,000 food processors, 113,000 food warehouses, and in excess of 1.2 million retail food facilities. Also, there are roughly 2.2 million farms, encompassing 920 million acres of land. Collectively, American farms produce $181 billion in crop production. The top five cash producing industries are grains and seeds, milk, poultry and eggs, fruits and nuts, and nurseries and greenhouses.

Beyond domestic food production, the Food and Agriculture Sector also imports many ingredients and finished products, leading to a complex web of growers, processors, suppliers, transporters, distributors, and consumers. Changes in the rules of trade, shifts in domestic policy, and new developments in technology have altered the competitive landscape of global agriculture and challenges facing American farmers. By providing food aid in disaster and poverty stricken areas around the world, these farmers also make a global humanitarian impact.

3. Sector-Specific Agencies

The Department of Health and Human Services' (HHS) Food and Drug Administration (FDA) and the United States Department of Agriculture (USDA) are designated as Sector-Specific Agencies (SSAs) for the Food and Agriculture Sector by Homeland Security Presidential Directive 7 (HSPD-7). The SSAs are responsible for leading sector-specific risk-reduction programs. USDA has responsibility for production agriculture and shares SSA responsibilities for food safety and defense with FDA. Specifically, FDA is responsible for the safety of 80% of all food consumed in the United States. The SSAs have been assigned responsibility for overseeing and coordinating protection and resiliency efforts.

3.1. USDA Leadership for SSA Responsibilities

USDA has statutory responsibilities to ensure plant and animal health and meat, poultry, catfish, frozen, dried, and liquid egg products, and catfish safety. USDA is also a research leader in human nutrition, animal and plant health protection, and new crop technologies that allow producers to grow more food and fiber using less water and pesticides.

USDA helps to ensure open markets for U.S. agricultural products and provides food aid to people in need domestically and overseas. USDA also provides a financial safety net to producers through market and disaster assistance programs and loans and a nutrition safety net for children and low-income people through the domestic nutrition assistance programs. Farming and ranching are the foundations of $1 trillion in food and fiber business with nearly $60 billion in annual exports. It generates almost 15% of the total economic activity in the nation, as well as providing nearly 18% of the country's jobs.

3.2. FDA Leadership for SSA Responsibilities

The FDA regulates $417 billion worth of domestic food, $49 billion of imported foods, and $60 billion of cosmetics sold across State lines. This regulation takes place from the products' point of U.S. entry or processing to their point of sale, with numerous food establishments (including food manufacturers, processors, and warehouses) and cosmetic firms. In addition, roughly 935,000 restaurants and institutional food service establishments and an estimated 114,000 supermarkets, grocery stores, and other food outlets are regulated by State and local authorities that receive guidance, model codes, and other technical assistance from FDA. FDA enhances its programs by supporting State and local authorities with training and guidance to ensure uniform coverage of food establishments and retailers.

The FDA also regulates animal feed. A safe animal feed supply helps to ensure the health of animals and people. To that end, FDA monitors and establishes standards for feed contaminants, approves safe feed additives, and manages the FDA's medicated feed and pet food programs.

The FDA primarily regulates food products sold in interstate commerce, whereas products made and sold entirely within a State are regulated by that State. In addition, formal agreements with the States for conducting inspections enhance FDA's ability to meet its public health mission. FDA personnel work with State agriculture and health departments to resolve food safety concerns and economic fraud cases.

Table 1 provides a summary of USDA and FDA jurisdictional overlap for food products.

Table 1. HHS/FDA and USDA/FSIS Jurisdictional Overlap for Commercial Food Products

Product	HHS/FDA	USDA/FSIS
Red meat products	Non-specified red meats (e.g., bison, rabbit, game animals, zoo animals, elk, wapiti, and moose)	Cattle, sheep, swine, goats, horses, mules, and other equine
Poultry	Non-specified birds: wild turkeys, wild ducks, and wild geese	Domesticated birds: chicken, turkey, ducks, geese, guineas, and ratites
Other meat products	Products containing <3% red meat (wet) and closed-faced meat sandwiches	Products containing 3% or more red meat (wet) and open-faced meat sandwiches
Other poultry products	Products containing <2% poultry (wet)	Products containing 2% or more poultry (wet)
Eggs	Shell eggs, products containing egg products and other egg processing not covered by USDA (e.g., restaurants, cake mix plants, and bakeries); Enforcement of shell egg labels/labeling	Pasteurized processed egg products, egg processing plants (washing, sorting, breaking, and pasteurizing)
Soup	All soup not covered by USDA	Soup containing 3% or more red meat or 2% or more poultry (e.g., chicken noodle)
Other products	Cheese, onion, mushroom pizza, spaghetti sauces (less than 3% red meat), spaghetti sauce with mushrooms and 2% meat, pork and beans, sliced egg sandwich (closed-faced), frozen fish dinner, rabbit stew, shrimp-flavored instant noodles, venison jerky, buffalo burgers, and alligator nuggets	Pepperoni pizza, meat lovers stuffed-crust pizza, meat sauces (3% or more red meat), spaghetti sauce with meatballs, open-faced roast beef sandwich, hot dogs, beef/vegetable pot pie, and chicken sandwich (open-faced)
Exceptions to the above	All foods involved in an outbreak aboard an interstate vessel, plane, train, or bus	–

3.3. CIKR Owners and Operators, Including Private and Public Entities

Regional and national organizations that represent the owners and operators have regular communication with the SSAs through conference calls and meetings to discuss protection and resiliency projects and initiatives underway by sector partners. SSAs continue to collaborate with these organizations on development and implementation of protection and resiliency strategies. The organizations can call on their members to provide additional knowledge and technical expertise across the full range of critical infrastructure protection activities and issues.

3.4. Department of Homeland Security

The Food and Agriculture Sector interacts with DHS' National Protection and Programs Directorate Homeland Infrastructure Threat and Risk Analysis Center that identifies and assesses current and future threats to the Nation's physical and informational infrastructure derived through the Strategic Homeland Infrastructure Risk Analysis process designed to assess and analyze, key risks to the Nation's CIKR, to include: terrorists, nation-states, malicious insiders, industrial accidents, lone-wolf assailants, and natural disasters. This information is communicated to the sectors via the Homeland Security Information Network.

4. Assessing Risk and Vulnerabilities

Operational risk management (ORM) was previously used within the sector to help prioritize food products/commodities for further assessments by evaluating relative public health consequences which yielded a risk ranking to facilitate decision making. The CARVER+Shock methodology was developed to assist with the assessment of vulnerability. CARVER+Shock was designed to identify vulnerabilities within assets, system, and networks that comprise the Food and Agriculture Sector by encompassing the consequences and threats.

4.1. CARVER+Shock

CARVER+Shock is an offensive targeting prioritization tool adapted from the military version for use in the food industry. The tool can be used to assess the vulnerabilities within a system or infrastructure to an attack. It allows the user to think like an attacker to identify the most attractive targets for an attack. By conducting a CARVER+Shock assessment of a food production facility or process, the user can determine the most vulnerable points in their infrastructure and focus resources on protecting the most susceptible points in their system.

CARVER is an acronym for the following six attributes used to evaluate the attractiveness of a target for attack:

- **C**riticality: Measure of public health and economic impacts of an attack
- **A**ccessibility: Ability to physically access and egress from target
- **R**ecuperability: Ability of system to recover from an attack
- **V**ulnerability: Ease of accomplishing attack
- **E**ffect: Amount of direct loss from an attack as measured by loss in production
- **R**ecognizability: Ease of identifying target

In addition, the CARVER tool evaluates a seventh attribute, the combined health, economic, and psychological impacts of an attack – the **Shock** attributes of a target.

The attractiveness of a target can then be ranked on a scale from one to ten on the basis of scales that have been developed for each of the seven attributes. Conditions that are associated with lower attractiveness (or lower vulnerability) are assigned lower values (e.g., 1 or 2), whereas, conditions associated with higher attractiveness as a target (or higher vulnerability) are assigned higher values (e.g., 9 or 10). Evaluating or scoring the various elements of the food sector infrastructure of interest for each of the CARVER + Shock attributes can help identify where, within that infrastructure, an attack is most likely to occur. Federal agencies, such as FSIS and FDA, have used this method to evaluate the potential vulnerabilities of farm-to-table supply chains of various food commodities. The method can also be used to assess the potential vulnerabilities of individual facilities or processes.

Federal agencies, such as the USDA/Food Safety and Inspection Service (FSIS) and the FDA, have used the CARVER + Shock method to evaluate the potential vulnerabilities of farm-to-table supply chains of various food commodities, as well as individual facilities or processes. These evaluations are carried out during face-to-face meetings of representatives from a particular segment of the food processing industry and Federal and State food safety agencies, and generally take 2–3 days. Using a scale from one to ten for each of the seven CARVER + Shock attributes, the participants score the "target attractiveness" of each segment, or "node", on a process flow diagram of the commodity or facility being evaluated. Conditions that are associated with lower attractiveness (or lower vulnerability) are assigned lower values (e.g., 1 or 2), whereas conditions associated with higher attractiveness (or higher vulnerability) are assigned higher values (e.g., 9 or 10). The individual scores for each CARVER + Shock attribute are then added together, so that each node in the diagram can have a total score ranging from 7 to 70.

Conducting face-to-face CARVER + Shock evaluations is resource-intensive and limiting in terms of the number of evaluations that can reasonably be conducted in any given time frame. Therefore, the FDA has sponsored development of CARVER + Shock software that can be downloaded. Having on-line CARVER + Shock software that produces results equivalent to those of a face-to-face session allows any member of the food processing industry to conduct a vulnerability assessment of their facilities and processes in a confidential manner.

The software tool is expected to be used by State and local food security agencies, industrial providers and any other parties interested in food defense. The tool is designed for use throughout the food industry. The pathogens considered are those that will survive most processes and therefore provide a worst-case scenario.

CARVER software mimics the thought processes in play during a face-to-face CARVER + Shock session by having the user:

Build a process flow diagram for the system to be evaluated.

Answer a series of questions for each of the seven CARVER + Shock attributes for each process flow diagram node.

Each question has an associated score. Based on the answers given, the software calculates a score for each CARVER + Shock attribute and sums them to produce a total score for each node. Analogous to a face-to-face session, total scores range from one to ten for each CARVER + Shock attribute and therefore 7–70 for each node. The user may view the attribute scores and total for each node, the total scores for all nodes, and the attribute scores for all nodes (e.g., all the node Criticality scores, Accessibility scores, etc.). The CARVER + Shock software can be downloaded at http://www.fda.gov/Food/FoodDefense/CARVER/default.htm.

5. Raising Awareness

5.1. ALERT

In 2006, FDA in cooperation with the Centers for Disease Control and Prevention, USDA, and State and local organizations representing food, public health, and agriculture interests announced a new food defense awareness initiative – ALERT. ALERT is an acronym that stands for Assure, Look, Employees, Reports, and Threat. It is linked to five questions designed to heighten awareness within the food sector on key food defense issues, such as product security and reporting of suspicious behavior. The ALERT initiative is intended to raise the awareness of state and local government agency and industry representatives regarding food defense issues and preparedness. It is generic enough to apply to all aspects of the farm-to-table supply chain and is designed to spark thought and discussion with a variety of stakeholders. ALERT identifies five key points that industry and businesses can use to decrease the risk of intentional food contamination at their facility.

The questions associated with the ALERT acronym are as follows:

A How do you **ASSURE** that the supplies and ingredients you use are from safe and secure sources?

L How do you **LOOK** after the security of the products and ingredients in your facility?

E What do you know about your **EMPLOYEES** and people coming in and out of your facility?

R Could you provide **REPORTS** about the security of your products while under your control?

T What do you do and whom do you notify if you have a **THREAT** or issue at your facility, including suspicious behavior?

ALERT initiative materials are available in English, Chinese, French, Korean, Portuguese, and Vietnamese. Additional information regarding the ALERT initiative is available on the FDA website at http://www.fda.gov/Food/FoodDefense/Training/ALERT/default.htm.

5.2. Employees FIRST

FIRST is a food defense awareness training kit for first line food industry employees. The training targets these individuals because they can play an important role in helping to keep our nation's food supply safe, from the farm to the table.

Food industry management can use the FIRST tool kit as part of ongoing employee food defense training programs. The tool kit focuses on five key points that industry and businesses can use to educate first line workers about the risks of food contamination. It also provides industry with measures to consider and implement to reduce these risks. Each of the letters in the FIRST acronym describes an action that a first line employee can take to mitigate risks of contamination.

F – **FOLLOW** company food defense plan and procedures
I – **INSPECT** your work area and surrounding areas
R – **RECOGNIZE** anything out of the ordinary
S – **SECURE** all ingredients, supplies and finished product
T – **TELL** management if you notice anything unusual or suspicious

Single copies of the kit are available in English and Spanish. The kit is free and includes one DVD, a training poster, and on-screen instructions. Additional information can be found on FDA's website at http://www.fda.gov/Food/FoodDefense/Training/ucm135038.htm.

6. Guidance Documents for Industry

6.1. Documents Issued by FDA

FDA has published the following four industry guidance documents on food security/defense. These documents identify the kinds of preventive measures that may be taken to minimize the risk that food will be subject to tampering or other malicious, criminal, or terrorist actions. The implementation of any of the measures included in each of these documents is entirely voluntary on the part of the industry. Notwithstanding, FDA's field personnel, as part of routine inspections, distribute and discuss these guidance documents with firms that have not previously received them.

FDA Food Producers, Processors, and Transporters: Food Security Preventive Measures Guidance, http://www.fda.gov/Food/GuidanceComplianceRegulatoryInformation/GuidanceDocuments/FoodDefenseandEmergencyResponse/ucm083075.htm.

FDA Importers and Filers: Food Security Preventive Measures Guidance, http://www.fda.gov/Food/GuidanceComplianceRegulatoryInformation/GuidanceDocuments/FoodDefenseandEmergencyResponse/ucm078978.htm.

FDA Retail Food Stores and Food Service Establishments: Food Security Preventive Measures Guidance, http://www.fda.gov/Food/GuidanceComplianceRegulatoryInformation/GuidanceDocuments/FoodDefenseandEmergencyResponse/ucm082751.htm.

FDA Dairy Farms, Bulk Milk Transporters, Bulk Milk Transfer Stations, and Fluid Milk Processors: Food Security Preventive Measures Guidance, http://www.fda.gov/Food/GuidanceComplianceRegulatoryInformation/GuidanceDocuments/FoodDefenseandEmergencyResponse/ucm083049.htm.

6.2. Documents Issued by USDA

Developing a Food Defense Plan for Meat and Poultry Processing Plants

A Food Defense Plan helps you identify steps you can take to minimize the risk that food products in your establishment will be intentionally contaminated or tampered with. http://www.fsis.usda.gov/pdf/food_defense_plan.pdf.

FSIS Model Food Security Plans for Import Establishments
http://www.fsis.usda.gov/PDF/Model_FoodSec_Plan_Import.pdf

Food Defense Plan – Security Measures for Food Defense
A model plan for small and very small establishments
http://origin-www.fsis.usda.gov/PDF/General-Food-Defense-Plan-9-3-09%20_2_.pdf

FSIS Safety and Security Guidelines for the Transportation and Distribution of Meat, Poultry, and Egg Products

Provides recommendations to ensure the security of food products through all phases of the distribution process. http://www.fsis.usda.gov/PDF/Transportation_Security_Guidelines.pdf

Food Defense Guidelines for Slaughter and Processing Establishments

Guidelines to assist federal and state inspected plants that slaughter and process meat, poultry, and egg products in identifying ways to strengthen their food protection. http://www.fsis.usda.gov/PDF/Securityguide.pdf.

7. Consumer Education on Food Tampering

Consumers play a critical role in preventing illness due to food tampering. FDA encourages consumers when shopping to carefully examine all food product packaging, check any anti-tampering devices on the packaging; not to purchase products if the packaging is open, torn, or damaged; not to buy products that are damaged or that look unusual; and to check the "sell by" dates. Consumers are also encouraged to carefully inspect products at home when opening the container and to never eat food from products that are damaged or that look unusual. FDA has created a fact sheet for consumer to assist them in identifying product tampering. The fact sheet, entitled "Food Tampering: An Extra Ounce of Caution," is posted on FDA's Web site at http://www.fda.gov/Food/ResourcesForYou/Consumers/ucm079137.htm.

8. Authorities from the Public Health Security and Bioterrorism Preparedness and Response Act of 2002 (Bioterrorism Act)

Section 303 of the Bioterrorism Act, Administrative Detention and Temporary Hold, authorizes FDA to order the administrative detention of food if an officer or qualified FDA employee finds credible evidence or information during an inspection, examination, or investigation that indicates the article presents a threat of serious adverse health consequences or death to humans or animals. This authority took effect upon enactment of the Bioterrorism Act in June 2002, and FDA issued final regulations implementing the procedures for exercising this authority in May 2004. Section 303 also authorizes temporary holds at ports of entry for a period not to exceed 24 h when FDA has credible evidence or information that an article of food presents a threat of serious adverse health consequences or death to humans or animals, or when FDA needs more time to inspect, examine, or investigate.

Section 304 of the Bioterrorism Act, Debarment for Repeated or Serious Food Import Violations, authorizes debarment of persons convicted of a felony for conduct related to the importation of any food or of persons who have engaged in a pattern of importing or offering for import adulterated food that presents a threat of serious adverse health consequences or death to humans or animals.

Section 305 of the Bioterrorism Act, Registration of Food Facilities, requires the owner, operator, or agent in charge of a domestic or foreign facility that manufactures, processes, packs, or holds food for consumption in the United States by humans or animals to register with FDA pursuant to this provision. The registration must contain the information necessary to notify the Secretary of HHS with the name and address of each facility at which, and all trade names under which, the registrant conducts business; the general food category as identified under 21 CFR 170.3; and for foreign facilities, the name and contact information of its U.S. agent. On October 10, 2003, FDA published an interim final rule to implement these provisions, which took effect on December 12, 2003 (68 FR 58894).

Section 306 of the Bioterrorism Act, Maintenance and Inspection of Records for Foods, provides that when FDA has a reasonable belief that an article of food is adulterated and presents a threat of serious adverse health consequences or death to humans or animals, persons (excluding farms and restaurants) who manufacture, process, pack, transport, distribute, receive, hold, or import food must provide access to records related to the food to assist FDA in determining whether the food is adulterated and presents a threat of serious adverse health consequences or death to humans or animals. This section also authorizes FDA to develop regulations that require the establishment and maintenance of records by persons (excluding farms and restaurants) who manufacture, process, pack, transport, distribute, receive, hold, or import food. Such records are to allow for the identification of the immediate previous sources and immediate subsequent recipients of food in order to address credible threats of serious adverse health consequences or death to humans or animals. FDA issued the final regulation implementing this section on December 6, 2004.

Section 307 of the Bioterrorism Act, Prior Notice of Imported Food Shipments, requires that FDA receive Prior Notice of food imported or offered for import into the US before the food arrives, which must include the article, manufacturer and shipper, grower (if known within the specified time in which notice is required), country of origin, country from which the article is shipped, and anticipated port of entry. On October 10, 2003, FDA published an interim final rule to implement these provisions, which took effect on December 12, 2003 (68 FR 58974). The purpose of Prior Notice is to better target efforts to monitor and inspect imported foods.

Section 308 of the Bioterrorism Act, Authority to Mark Articles Refused Admission into the US, authorizes FDA to require the marking of refused food (other than food required to be destroyed). Marking is to be done at owner's expense. This provision also makes food misbranded if it fails to bear the required label when FDA has found the food presents a threat of serious adverse health consequences or death to humans and animals and FDA has notified the owner or consignee that the label is required and that the food presents such a threat.

Food Safety Strategies in the Federal Republic of Germany

Joachim Gaus

Federal Ministry of Food, Agriculture and Consumer Protection, Wilhelmstraße 54, 10117 Berlin, Germany

Abstract

Food regulation is essentially harmonised in the European Community (EC). National provisions exist only where Community law leaves regulatory gaps or where national specifications are required for the implementation of Community law. Community and national legal provisions provide for a high and, at the same time, non-discriminatory level of protection in the area of food safety. Only safe food may be marketed, irrespective of whether it comes from Germany, an EC Member State or from abroad – a so-called third country.

1. Organisation of Official Food Control and Inspection in Germany

The Federal Republic of Germany is a federal republic consisting of 16 federal states, or *Länder*. The general distinction between competences of the Federal Government and the *Länder* in the area of food and feed safety, animal health, animal welfare and plant health is laid down in the Basic Law of the Federal Republic of Germany from 1949.

Thus, official food control and inspection in Germany lies within the competence of the *Länder*. In assuming its tasks, official food control and inspection authorities have recourse to a coherent and multi-layered legal framework which has been developed by the European Union (EU) in recent years. The work of the EU Commission in cooperation with the Members States aims at further harmonisation.

The objective of implementing a horizontal approach in the field of food safety within the Community is laid down in Regulation (EC) No 178/2002[1] – the so-called basic regulation. With this basic regulation, the Community has adopted the

[1] Regulation (EC) No 178/2002 of the European Parliament and of the Council of 28 January 2002 laying down the general principles and requirements of food law, establishing the European Food Safety Authority and laying down procedures in matters of food safety (OJ EC No L 31, p. 1).

H. Alpas and B. Cirakoglu (eds.), *Food Chain Security*,
DOI 10.1007/978-90-481-9558-9_8, © Springer Science + Business Media B.V. 2010

idea of the value-added chain. The Regulation looks at all aspects of food production as a continuum, from primary production – feed, vegetable and animal products – all the way to the delivery of food to consumers.

The basic regulation clearly states that food operators are responsible for the safety of their products. Food operators must ensure that their foodstuffs meet the requirements of food law at each stage, from production to processing and marketing. Furthermore, food operators must set up systems to ensure traceability. This means that they must document when and from whom they purchased which product and whom they sold it to.

Food control and inspection authorities are entitled to receive information from operators' internal traceability systems. This enables monitoring systems to carry out corresponding control measures to prevent further food safety problems.

Moreover, the basic regulation also contains rules on crisis management, the rapid alert system for food and feed as well as on managing food chain emergencies. These rules are designed to provide additional information to food control and inspection authorities and to strengthen their capacity to act, particularly in cooperation with European neighbours.

By adopting Regulation (EC) No 882/2004[2] – the so-called control regulation – the EU has further developed the horizontal approach, which it first introduced with the basic regulation. The control regulation aims specifically at authorities in charge of monitoring. It contains guidelines for the control of food and feed law provisions and for the verification of compliance with animal health and animal welfare provisions. A major aspect of this regulation is that official inspections must be conducted in a risk-based manner. Moreover, the control regulation stipulates that there must be effective coordination between different inspection authorities.

In terms of normative, national measures, the Federal Government has adopted the Community-wide horizontal regulatory concept by enacting the Food and Feed Code,[3] which brought together provisions on food, commodities and feedstuffs in a single legal instrument.

Federalism's decentralised system offers several advantages for food control and inspection, e.g. a proximity to the production and processing sectors. This should be taken advantage of in the best possible way. A prompt exchange of information among inspection authorities and effective coordination for different competences are of essential importance. For this purpose, by enacting the general administrative regulation (AVV) on framework monitoring in 2004, the Federal Government created the conditions for better coordination and nation-wide harmonisation of

---

[2] Regulation (EC) No 882/2004 of the European Parliament and of the Council of 29 April 2004 on official controls performed to ensure the verification of compliance with feed and food law, animal health and animal welfare rules (OJ EC No L 191, p. 1).

[3] Notice of the recast version of the Food and Feed Code as amended by the notice of 24 July 2009 (Federal Law Gazette I p. 2205).

food control and inspection. This provision, which has been revised in the meantime, aims at guaranteeing a high level of consumer protection nation-wide.[4]

2. Competent Authorities and General Division of Tasks

2.1. At Federal Level

Under the German Food and Feed Code, almost all tasks in the areas of food and feed safety, animal health, animal welfare and plant health, which are to be performed at federal level, lie within the competence of the Federal Ministry of Food, Agriculture and Consumer Protection (*BMELV*). To fulfil the tasks within its sphere of responsibilities, the ministry works with a number of other bodies, including Superior Federal Authorities, legally independent public institutions and federal research centres. Certain tasks in the field of food safety lie within the competence of the Federal Ministry for the Environment, Nature Conservation and Nuclear Safety (*BMU*) and the Federal Ministry of Justice (*BMJ*).

2.1.1. Federal Ministry of Food, Agriculture and Consumer Protection (BMELV)

The *BMELV's* main competences comprise first and foremost the protection of animal health, ensuring that food is safe, the protection of plant health, promoting the economic viability of farm holdings and supporting the development of ecologically and socially unspoilt rural areas. In Germany, the *BMELV* is in charge of drafting legislative initiatives and general administrative regulations and of enacting regulations in this area. This also includes a multiannual control plan for official inspections in the food and feed sector.

2.1.2. Federal Ministry for the Environment, Nature Conservation and Nuclear Safety (*BMU*)

Preservation of the environment is another important mainstay for food safety. The objective of environmental policy, with respect to environmental contaminants, is to create the right conditions for producing healthy foodstuffs that are as uncontaminated as possible. Within the Federal Government, the *BMU* is in charge

[4] General Administrative Regulation of 21 December 2004 on guidelines for the implementation of official monitoring performed to ensure the verification of compliance with feed, food wine and tobacco law provisions (Joint Ministerial Gazette No 58 p. 1161), as last amended by General Administrative Regulation of 3 July 2008 (Joint Ministerial Gazette No 22 p. 425).

of protecting foodstuffs from hazards from the air, water and soil. The *BMU* monitors environmental contaminants such as heavy metals (cadmium, lead and mercury), dioxin and PCB.

2.1.3. Federal Office of Consumer Protection and Food Safety (*BVL*)

The *BVL* works in close cooperation with the *BMELV* in the area of food safety, veterinary matters and consumer protection and with the *BMU* on matters of food safety. The *BVL* provides data and technical assistance. A key focus of its work is risk management.

The *BVL* acts as inspection authority for the authorisation of novel food and as authorising authority for foodstuffs which do not meet Germany's food law provisions. Furthermore, the office is also responsible for authorising plant protection products for the German market and for the control and monitoring of active ingredients within Europe. In addition, the *BVL* is also in charge of authorising veterinary drugs in Germany. The *BVL* also evaluates applications for authorisation on the European market.

The *BVL* ensures a flow of information between the Federal Government and the federal states (*Länder*) in the areas of food safety, veterinary medicine and consumer protection. Furthermore, it acts as the main office for the working group of food chemistry experts of the *Länder* and the *BVL* (*ALS*) and for the working group of experts on food hygiene and foods of animal origin (*ALTS*). It also assists in drafting general administrative regulations and coordinates the development and implementation of nation-wide monitoring, surveillance and control schemes.

In accordance with the German Food and Feed Code, the provisional German Tobacco Act and the German Genetic Engineering Act, the *BVL* publishes an official list of sampling and control procedures for food, cosmetic products, commodities, tobacco products and animal feed and also of sampling procedures for monitoring work at federal level within the framework of the Genetic Engineering Act. The *BVL* then has the task to use this as a basis for standardising sampling and testing methods.

The *BVL* is the national contact point in the EU Rapid Alert System for Food and Feed (RASSF) and the central contact point for the European Commission's Food and Veterinary Office. Moreover, the *BVL* is also home to the European and national reference laboratory for residues of pharmacologically active substances and contaminants in food of animal origin, and to a number of other national reference laboratories (see Section 2.4).

In crisis situations, the unit within the *BVL* which is in charge of crisis management supports the relevant unit within the *BMELV*. Moreover, the *BVL* also provides logistical and organisational support for the *Länder*.

The founding of the *BVL* strengthened the cooperation between the Federal Government and *Länder* authorities and the ministries and experts in all relevant areas. This makes it possible to promptly react to new problems and crisis situations and allows for differentiated, cross-border action within the framework of EU programmes.

2.1.4. Federal Institute for Risk Assessment (*BfR*)

The *BfR* is a scientific agency which prepares expert reports and opinions on matters of food safety and consumer protection on the basis of internationally recognised scientific evaluation criteria. The *BfR* enjoys independence with respect to its scientific opinions.

The *BfR* compiles scientific opinions and reports on food and feed safety, including questions of diet and dietary prevention. Furthermore, it assesses health risks and the safety of use of chemicals, plant protection products and pesticides. Further duties include documenting cases of, and providing information about, intoxications, documenting and evaluating alternative methods to animal experiments and conducting risk assessments for genetically modified organisms and novel foods.

Based on the results of risk analysis, the *BfR* formulates management options designed to reduce risks. The evaluations are presented to the public, the scientific community and other stakeholders and interested parties in a transparent and easily comprehensible manner. Evaluation results are made accessible to the public whilst respecting the confidentiality of protected data.

The *BfR* provides information and advice about its entire field of activity to federal ministries and informs authorities of the *Länder*, EU institutions, the business community and the public at large about its findings.

The federal institute works in close cooperation with other scientific institutions at national and international level and with institutions of other countries which are active in the field of consumer health protection and food safety. A key focus is its cooperation with the European Food Safety Authority (EFSA).

The *BfR* plays an active role in conducting food monitoring and national surveys in the area of feedstuffs and feed additives and in the documentation and evaluation of trends in zoonotic agents and of sources of zoonotic infections.

An important task of the *BfR* is to make information available to the public and to engage in a participative risk communication dialogue with the different interest groups.

The *BfR* conducts its own research on topics which are closely related to its own tasks in the areas of consumer health protection and food and feed safety.

National reference laboratories are attached to the *BfR* (see Section 2.4).

2.1.5. Friedrich–Loeffler-Institute, Federal Research Institute for Animal Health (*FLI*)

The *FLI* is an independent higher federal authority affiliated with the Federal Ministry of Food, Agriculture and Consumer Protection (*BMELV*). The *FLI* advises and assists the *BMELV* on all matters related to animal health. To this end, the *FLI* performs special tasks with respect to monitoring and research of animal disease, primarily of farm animals (cattle, pigs, poultry and fish). The tasks include

- Research on animal diseases, including zoonoses, animal nutrition, animal husbandry, animal welfare and livestock genetics
- Examination of animals and animal products intended for import or export
- Epidemiological studies in the event of animal epidemic outbreaks

In addition, the *FLI* also functions as the national reference laboratory for compulsorily notifiable animal diseases and Q fever and performs functions at federal level with respect to the quality management of laboratory diagnostics regarding compulsorily notifiable animal diseases. More than 40 national reference laboratories, a WHO Collaborating Centre (for rabies) and five OIE reference laboratories are currently attached to the *FLI*. Alongside their official and sovereign mandates, all reference laboratories also conduct application-oriented research.

Another of the *FLI's* tasks is to monitor the occurrence of endemic and newly introduced animal diseases. To this end, the Institute of Epidemiology manages the German animal disease notification system (*TSN*) as the official reporting system for compulsorily notifiable animal epidemics and reportable animal diseases. The *FLI* also participates in the Trade Control and Expert System (TRACES), an EU project which is designed to document transport of livestock and products of animal origin within the EU and from third countries.

By conducting research on epidemiological links and the development of bio-mathematical models, the *FLI* also participates in the risk assessment of animal diseases. Another of the *FLI's* tasks is to regularly update the official list of sampling and control procedures regarding samples of animal origin for compulsorily notifiable animal diseases (the so-called "method compilation"). This task is binding for the competent authorities and inspection agencies.

2.1.6. Julius Kühn-Institute (*JKI*)

The *JKI* conducts research on plant health and plant protection. Its main task is to advise the *BMELV* on matters of plant protection and plant health as well as on biosecurity. As a research institute, it conducts research on all aspects of plant protection and plant health. The institute offers technical information in the area of plant production and risk analyses and participates in the development of national and international health standards for plants.

2.1.7. Max Rubner-Institute (*MRI*)

The *MRI* is the *BMELV's* research and advisory centre and focuses on research on consumer health protection in the food sector. Other important tasks include in particular the identification and nutritional–physiological assessment of health-relevant food ingredients, the examination of gentle, resource-conserving methods of processing and handling, quality assurance of foods of plant and animal origin as well as the investigation of dietary habits and the improvement of consumer information. Two national reference laboratories are attached to the *MRI*.

2.2. Administrative Bodies in the Länder

As, according to the German Basic Law, implementation of laws lies within the competence of the *Länder*, the different federal ministries coordinate the monitoring of the individual inspection authorities in the field of veterinary matters, food and feedstuffs in the respective *Land*.

At *Land* level, veterinary and food administration is divided into up to three subdivisions.

2.2.1. Supreme Land Level

The supreme *Land* authority is the Ministry/Senate Department responsible for food, feedstuffs and veterinary matters; its tasks include monitoring, planning, managing and coordinating, and issuing instructions regarding, all areas of the respective *Land* relating to food, feed and veterinary matters.

In some *Länder*, the tasks of food and feed monitoring and, in some cases, of animal welfare and plant protection are implemented by higher *Land* authorities.

2.2.2. Intermediate Land Level

Some *Länder* are divided into administrative regions. These regions are administered by regional administrations or by district governments which constitute the intermediate administrative level between the supreme *Land* authority for food and feed monitoring and veterinary matters and the monitoring authorities of the administrative districts and urban municipalities. An intermediate level such as this exists in five *Länder*.

Regional administrations and district governments are controlled respectively by the supreme food and feed inspection authority and veterinary authority within the framework of specialized supervision (audit).

2.2.3. Lower Land Level

At district level (lower administrative level), monitoring is conducted by food inspection and veterinary authorities of the administrative districts and urban municipalities. There are approximately 400 authorities at this level in Germany. These authorities are controlled by the respective competent supreme or intermediate food and feed inspection authority and veterinary authority within the framework of specialized supervision (audit).

2.2.4. Inspection Authorities

Moreover, there are 41 state-run chemical and veterinary investigation centres as well as further municipal testing authorities which are under the control of the *Länder*. In addition, five *Länder* have agricultural institutes which conduct laboratory tests.

2.2.5. Audit Schemes of the Competent Authorities

The *Länder* are responsible for conducting audits in accordance with the control regulation. They determine the type of audit procedure to be conducted, the planning cycle, follow-up and proof of transparency. Within the framework of the *Länder* working group on consumer protection (*LAV*), a special working group on quality management has been set up in order to create the general requirements for the implementation of the control regulation. A harmonised approach of the *Länder* for the internal auditing of authorities has been developed and adopted by the *LVA*.

2.2.6. Coordination and Communication Between the Federal Government and the Länder

Coordination and communication between the Federal Government and the *Länder* is ensured in several ways. The BMELV organises ad hoc sessions with the ministries of the *Länder* to provide advice on questions of law-making. The

Länder have also developed coordination and communication structures (working groups which act as interface between the federal and *Länder* level) in order to bring about a nationwide, harmonised implementation of the relevant legislation. Panels at political, strategic and working level have been set up to support coordination and communication.

A standing conference of Ministers of Consumer Protection (at political level) was set up in order to support changes in legislation and to introduce new enforcement procedure for all *Länder*. The conference brings together senior officials from the involved ministries of the *Länder* and the Federal Ministry. The conference meets at least once a year and deals with matters of health and economic protection of consumers. The chair of the conference changes annually. The Federal Government participates in these meetings and has the right to vote (except for questions which only affect *Länder* matters). As a result of the meetings of this conference, inquiries are being addressed to the *Länder* working groups on consumer protection for further examination. The consumer protection authorities of the *Länder* are cooperating in the *LAV*. The *LAV's* task is primarily to coordinate the implementation of legal provisions. To this end, it has set up 11 working groups:

- Animal diseases, animal health (AGTT)
- Animal welfare (AGT)
- Animal feed (AFU)
- Food and commodities, wine and cosmetics (ALB)
- Meat and poultry meat hygiene, technical questions about food of animal origin (AFFL)
- Veterinary medicinal products (AGTAM)
- Healthy diet and nutritional information (AG GEE)
- Import, export and transit (AG EAD)
- Economic consumer protection (AGWV)
- Educational and professional matters of persons working in the field of food and veterinary law (AFAB)
- Information and Communication (AG IuK)
- Quality management on consumer protection

The *LAV* members comprise the heads of the consumer protection departments of the *Länder*. If required, representatives of federal ministries and other authorities, organisations and professional associations or scientific institutions may also be invited to the meetings. Each working group is chaired by a *Land*. The chair changes every 2 years. The working group formulates recommendations and suggestions based on unanimity which are then submitted to the Conference of Consumer Protection Ministers. Upon deciding on a recommendation, the individual *Länder* are bound to apply the agreed rules. In order to deal with specific topics in a timely manner, the *LAV* or its working groups may set up project groups which will then submit a final report for decision-taking.

2.3. National Reference Laboratories

Pursuant to the control regulation, Germany has set up the following reference laboratories in the areas of food, feed, animal health and plant health:

- The *BfR* conducts studies and tests on zoonoses (salmonella, *Listeria monocytogenes*) and monitors marine biotoxins, viruses and bacteria in bivalve molluscs, coagulase-positive staphylococci including *Staphylococcus aureus*, *Escherichia coli*, verotoxin-producing *E. coli*, Campylobacter, parasites (particularly trichinella), resistance to antibiotics, animal protein in feedstuffs, additives in animal feed, food contact materials, mycotoxins in food and feed, dioxin and PCB in food and feed.The *BVL* conducts studies and tests on residues of plant protection products (cereals and food, feed, food of animal origin and products with high fat content, fruits and vegetables including products with high water and acidity content), genetically modified organisms, polycyclic aromatic hydrocarbons, heavy metals in food and feed, residues of veterinary medicinal products and harmful substances in food of animal origin.
- The *MRI* conducts studies and tests in the areas of milk, milk products and anisakis.
- The *FLI* conducts studies and tests for: African swine fever, avian influenza, brucellosis, echinococciasis, exotic animal disease (African horse sickness, bluetongue disease), transmissible spongiform encephalopathies (TSE), classical swine fever, Newcastle disease, swine vesicular disease, fish diseases, molluscs and crustaceans diseases, rabies, foot and mouth disease, infectious haematopoietic necrosis and viral haemorrhagic septicaemia. Moreover, the *FLI* also manages a range of additional national reference laboratories for compulsorily notifiable animal diseases and for animal diseases that are reportable under German law.
- The *JKI* is the reference laboratory for plant health – in particular for EU-wide monitored organisms – and with respect to directives on the control of harmful organisms (e.g. *Clavibacter michiganensis* subsp. *sepedonicus* and *Bursaphelenchus xylophilus*).

The 16 federal *Länder* designate laboratories to conduct the official food and feed monitoring programme as well as animal and plant health checks.

Assessing Security of Supply: Three Methods Used in Finland

Hannu Sivonen

National Emergency Supply Agency, Pohjoinen Makasiinikatu 7 A, FI-00130 Helsinki, e-mail: Finland, hannu.sivonen@nesa.fi

Abstract

Public Private Partnership (PPP) has an important role in securing supply in Finland. Three methods are used in assessing the level of security of supply. First, in national expert groups, a linear mathematical model has been used. The model is based on interdependency estimates. It ranks societal functions or its more detailed components, such as items in the food supply chain, according to the effect and risk pertinent to the interdependencies. Second, the security of supply is assessed in industrial branch committees (clusters and pools) in the form of indicators. The level of security of supply is assessed against five generic factors (dimension 1) and tens of business branch specific functions (dimension 2). Third, in two thousand individual critical companies, the maturity of operational continuity management is assessed using Capability Maturity Model (CMM) in an extranet application. The pool committees and authorities obtain an anonymous summary. The assessments are used in allocating efforts for securing supply. The efforts may be new instructions, training, exercising, and in some cases, investment and regulation.

1. Security of Supply in Finland

1.1. Security of Supply in Legislation

In the Finnish legislation, security of supply is defined as the security of the livelihood of the population, continuity of vital economic activity and functioning of the infrastructure in normal conditions, as well as during serious disturbances, and in exceptional circumstances (Laki huoltovarmuuden turvaamisesta 18.12.1992/1390). The markets usually provide security of supply, but special arrangements are sometimes needed (National Emergency Supply Agency 2009a).

In the most recent government decision on the targets of security of supply (Government decision on the targets of security of supply 21.8.2008/539) the

H. Alpas and B. Cirakoglu (eds.), *Food Chain Security*,
DOI 10.1007/978-90-481-9558-9_9, © Springer Science + Business Media B.V. 2010

security of supply is divided into two main areas: security of critical infrastructure and security of critical production. Both of these areas must be secured in order to make sure that the people get food, that their houses are heated and provided with electricity and communication, that they obtain health care, and that they have all other products and services necessary for life. Foreign trade plays an important role in this for Finland (Figure 1).

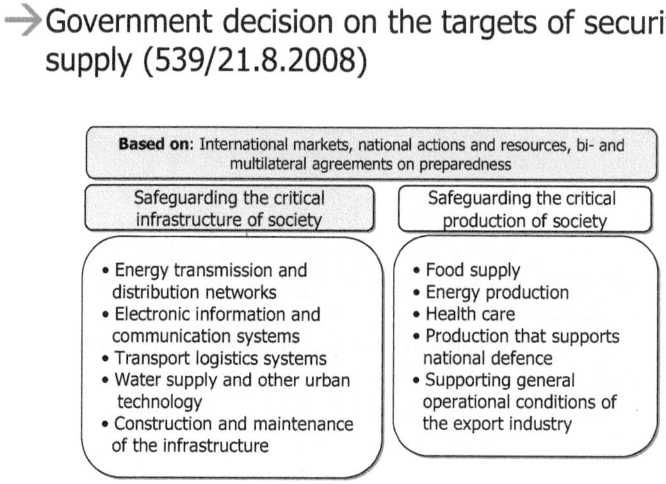

Figure 1. Targets of security of supply

1.2. Food Supply

Food supply is one of the critical production areas. The food chain consists of primary production, food processing industry, and the logistic chains needed to transport the necessary raw materials and food components to farms and industry, as well as the distribution chain to the consumer. The food chain uses critical infrastructures: energy networks, information and financial systems, transport logistics, and water systems.

1.3. Public Private Partnership (PPP) as a Means to Secure Supply

In order to secure the supply, there are several tools available for the government, the authorities, and the industrial actors:

International agreements, legislation and possible regulation provide the framework for security of supply. The government's economic and industrial policy takes into consideration the point of view of security of supply.

A vital tool since 1955 is the public private partnership organisation, which now consists of about 1,000 leading experts in 24 critical industrial areas. The partnership committee organisation, organised as cluster and pool committees, exchanges information within sectors and across sectors, follows the business environment and threats to it, supports individual operational continuity management, arranges exercises, and carries out surveys and research and development projects with the help of consultants and the academia (Figure 2).

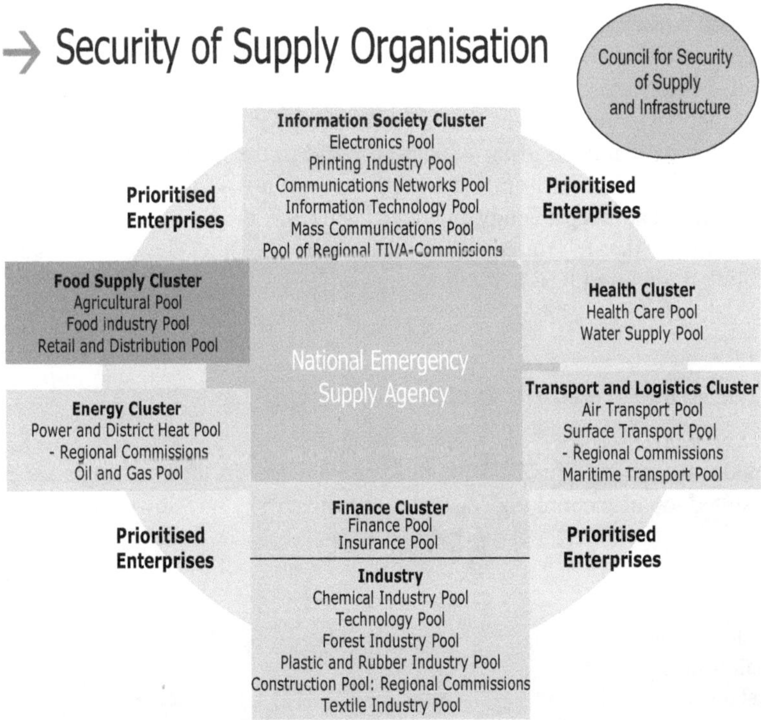

Figure 2. Security of Supply Organisation in Finland

The high level Council for Security of Supply and Infrastructure (CSSI) gives guidelines to the Security of Supply Organisation. National Emergency Supply Agency has a central role in the Security of Supply Organisation. An Agency representative is a member in each cluster and pool committee. The Agency has a fund which can be used in selected cases to finance security of supply measures. Important objects of financing are the fuel, grain and seed grain, and medical supply stockpiles.

The assessment of security of supply is done on three levels: national level, in the industrial area level in the 24 industrial pool committees, and in the individual actor or company level.

2. Risk Assessment Based on Interdependencies

2.1. Linear Model

The first method described here is a linear mathematical model used to rank the risks involved in different societal or technical functions and systems, when the interdependency of these functions is taken into consideration (Sivonen 2005a).

As all models, this one is not an exact picture of the real world, but it helps an expert group to structure their discussion and to reach a consensus. The result of the risk ranking can be used for allocating effort to the most risky areas. Thus the risk ranking in 2005 clearly raised ICT to the top in risk ranking and encouraged deeper study. The deeper study included using the model again in 2006, but in more detail in the ICT area. After that, also measures have been taken in controlling the risks in ICT.

2.2. The Model Used for Food Chain

The model was piloted in 2004 for ranking elements in the food logistics chain. The resulted top of the ranking was, from the most risky one downwards:

Logistics centres
Order systems
Cashier systems
Wholesale companies
Retail outlets
Meat operators
Dairy operators
Grain and vegetable operators
Medium-sized retail outlets
Hypermarkets
Cooling equipment

The rest of the 20 components in the logistics chain were ranked clearly less risky.

2.3. Details of the Model

The model defines dependency of function f_1 on function f_2 as the share a_{12} that the failures in f_2 have as a cause in the (consequences of) failures in f_1.

Let us consider a very simple system of these two functions f_1 and f_2. The dependencies in this system would be:

$$a_{11}\ a_{12},\ \text{where } a_{11} + a_{12} = 1$$
$$a_{21}\ a_{22},\ \text{where } a_{21} + a_{22} = 1,$$

in other words each of the *horizontal* rows cover all causes (1 = 100%) for a function to fail.

The items a_{11} and a_{22} represent the share of internal failure causes. For instance, failure in electricity may be caused by the failure of a technical component in the electricity system, or an operating failure of the electricity system, as opposed to a failure in an external function such as data communication. The model takes into account both alternatives, internal and external causes of failure.

Let c_1 and c_2 be the consequences for f_1 and f_2 failing respectively for one unit of time. The consequences can be e.g. costs in monetary terms, or some other measurable consequences.

Now let $c^d_1 = a_{11}c_1 + a_{21}c_2$ and
$c^d_2 = a_{12}c_1 + a_{22}c_2$.

Here c^d_1 and c^d_2 are the derived consequences of f_1 and f_2 failing respectively, when also the interdependencies are taken into consideration. The "guilt share" is added up *vertically* to represent the indirect effect of the failures of each function f_1 and f_2. (See Appendix at the end of this article. It explains the method in more detail).

If one replaces c^d_1 and c^d_2 for c_1 and c_2 and repeats the same calculation over and over again, the c^d_1 and c^d_2 converge in all practical cases into a so called dominant eigenvector of the matrix a_{ij}. This was proved by the German mathematician Carl Jacobi in 1846 (Mathews 2004a, b). The repetition renders visible the effects of all indirect dependencies which are contained in the matrix a_{ij}.

Fortunately, the initial direct consequences c_i need not be estimated. Arbitrary figures can be used, because the matrix a_{ij} defines completely the proportional values of c^d_i, and therefore also the ranking.

The linearity assumption is a simplification. The shares a_{ij} in real world are not constant, but they change case by case and over time. However, this simplification is intuitively appealing and seems to make sense. Also, the mathematical consequences of this linear model seem to make sense. The most important consequence is that the functions can be ranked using the model.

2.4. Estimating Interdependencies

In order to facilitate the estimating of the coefficients a_{ij} a rough scale was used (Figure 3):

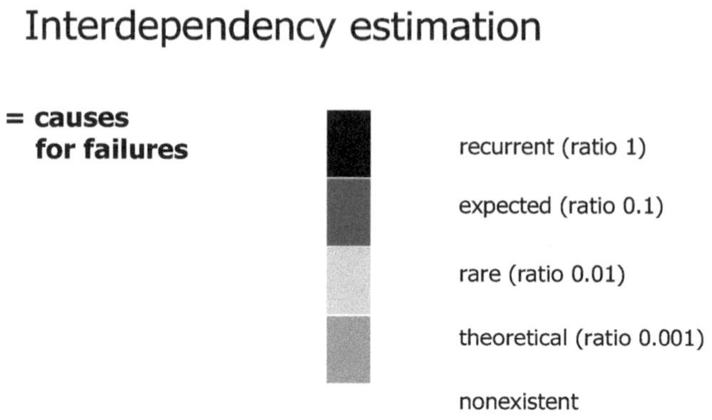

Figure 3. The rough scale of interdependency estimation

When failure in one function f_j is recurrently the reason of failure in another f_i, we used $a_{ij} = 1$; when it is expected, we used 0.1; when it is rarely a reason; we used 0.001. After this preliminary estimation, we scaled the whole row of a_{ij} to sum up to 1, i.e. then all reasons for the failures of f_i are then covered 100%. The use of this rough preliminary scale with tenfold steps did not seem to have a great effect on the resulting ranking. Approximately the same ranking results from other scales, too.

The estimation of dependencies is the most important step using the model. It gives the basic ranking.

In Figure 4, one can see that even circular dependencies may exist. Transport management systems are dependent on electricity, end-to-end transport chain is naturally dependent on transport management, and finally, failures in transport chain may in some cases prolong failures in electricity systems, if manpower and spare parts cannot be transported to the point of failure in time.

The model that was used in ranking the critical infrastructures and critical production in 2005 had 60 items (functions and threats) to be considered. Therefore, there were 60 * 60 = 3,600 dependency estimates to think of. The simplified estimation method, described above, facilitated the estimation. It was also noticed, that e.g. for a data communication expert it is easy to estimate the reasons why data communication fails, whereas it would be very difficult to think the other way round: what is the relative effect of data communication failures in other fields.

In fact, for later studies in ICT area, six different (3 * 2) scenarios were constructed. They consisted of failures of different durations (hours, days, and months in extreme situations). Another factor was if the failure of one function was the cause of a failure in another, or if it was an obstacle for the recovery of another. In this manner, six slightly different dependency matrices were built and six slightly different rankings were obtained. This approach gave a better insight to the risks in ICT.

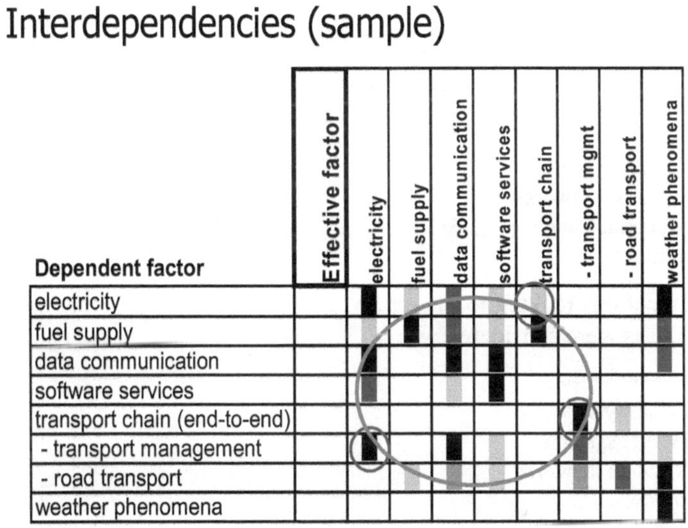

Figure 4. A sample of the 60 * 60 matrix of interdependencies of critical functions and threats in Finland

2.5. Estimating Mean Times Between Failures and Durations of Failures

The basic model with interdependency estimates may be completed with an estimation of failure probabilities of each function. The probabilities are derived from estimated mean times between failures and the durations of failures, based on the assumptions of Poisson process (Nenadic 2002). Simple and rough classification is used here, as in the estimation of interdependencies.

The Poisson process is the simplest model for random events which occur independently. The ranking obtained, when the probabilities are used, may be slightly different from the basic one. After some experience with the model, it has turned out that the basic ranking usually is good enough, because the dependency estimates already contain an idea of the frequency of failure occurrences.

2.6. Ranking Results for Critical Infrastructures, Critical Production, and Threats to Them

The functions ranking to the top in 2005 out of 43 functions, when only the interdependencies were considered, were

Electricity
Software services
Data security services
Data server systems
Work station networks
Data communication
Air traffic

The functions ranking to the top, when also the failure probabilities and therefore the risks were considered, were

Software services
Data security services
Data server systems
Work station networks
Electricity
Air traffic
Data communication

The top threats out of the 17 threats considered, were

Weather phenomena (storms, frost, flooding)
Threats to information systems (malware, denial of service attacks, etc.)
Crisis in international logistics
Fire or water damage in buildings
Crime and terrorism

3. Security of Supply Indicators

3.1. Definition of the Indicators

A second method is used in the PPP committee (cluster and pool) level. Every 2 years each business branch evaluates the overall security of supply in their area. Expert knowledge is condensed into *Security of supply indicators* (Sivonen 2005b).

The indicators are assessments on a scale *good*, *satisfactory*, *tolerable*, *and bad*. Also, if a trend is seen, it is indicated. Simple traffic light colour symbols are used (*green*, *amber*, *red*, *black*).

The same five factors are considered in each of the committees (Figure 5):

Two redundancy factors: (1) capacity and (2) alternatives available, and three other factors: (3) controllability from Finland, (4) security arrangements, and (5) preparedness. These are expressed as vertical columns in a table.

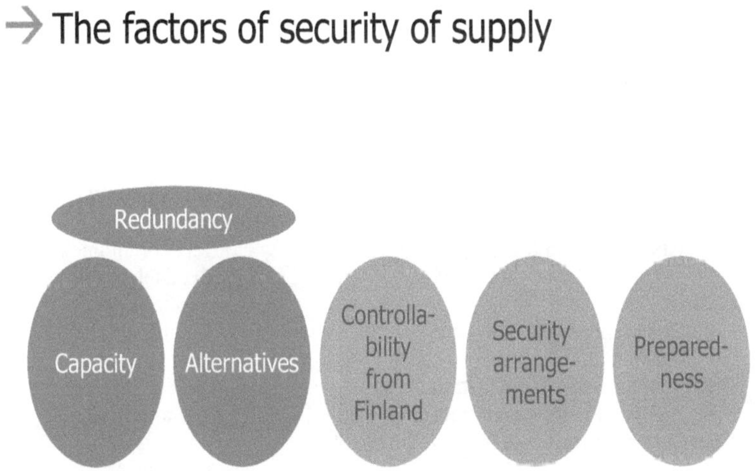

Figure 5. The factors of security of supply

About factor (2), alternatives available: The existence of several parallel food supply chains is considered to be a positive thing for the security of supply. In case one of them is disturbed for any reason, the others can take at least part of the load. Factor (3), controllability from Finland, can mean self sufficiency or Finnish majority ownership. Factor (4), security arrangements, includes physical protection and guarding of facilities, adequate security procedures, and also cyber security measures. Factor (5), preparedness, includes operational continuity management in the companies, contingency planning in the branch, exercises, and awareness of changes in business environment and threats.

Each branch defines the horizontal lines of the indicator table: the specific functions in the business that contribute to security of supply.

3.2. Functions Considered in Food Supply

In food supply the following taxonomy is used:

Food supply system
Food safety

Primary production
- Plant production
  - Grain production
  - Grass production
  - Oil plant production
  - Sugar beet production
  - Potato production
  - Other plant production
- Domestic animal production
  - Milk production
  - Pork production
  - Beef production
  - Poultry production
  - Fishing and fish breeding

Processing
- Grain based industry (incl. feed industry)
- Drink and nutrition fat industry
- Meat, fish, and ready meal industry
- Vegetable based and sweetener industry

Logistics and distribution
- Central warehouses, logistics centres
- Distribution centres, distribution terminals
- Hotel, restaurant, and catering operators
- Distribution system in cities
- Distribution system in towns
- Distribution system in sparsely populated areas
- Distribution transport

Input production and indispensable import
- Grain import (rye)
- Feed and feed raw materials import
- Pesticide import
- Fertilizer import
- Fertilizer production
- Meat import (beef)
- Prepared food import

Administrative preparedness of food supply system (in case i.e. rationing is needed)

Support functions for food supply
- Energy supply
  - Electricity supply
  - Fuel supply
  - Heating
- Information systems
  - Order- and delivery systems
  - Production control systems
  - Administrative systems
- Communications
  - Data communication
  - Fixed telephone network
  - Mobile telephone network
  - E-mail
  - Mail
- Transport subcontractors
  - Forwarding
  - Sea transport
  - Air transport
    - Refrigerated transport
  - Land transport
    - Refrigerated transport
- Maintenance
- Water supply and sewage
- Waste management
- Financial services
  - Payment systems
- Insurance services
- Mass media

There are 67 horizontal lines in the table. Thus there are 5 × 67 = 380 indicators to be considered. No straightforward arithmetic can be used when assessing the indicator values of *Processing*, for instance, from its component subfunctions, but they must be assessed individually.

The purpose of the indicators

The indicator approach provides a framework for assessment discussion in an expert group. In the discussion, many phenomena in the business environment, economy, and technology must be considered. The result is a situation picture combined with textual report. This will be used for allocating effort to training, planning, exercising, and perhaps some special investment with or without financial support from National Emergency Supply Agency.

The indicator method has been used since mid-1990s and updated in 2006.

4. Maturity Analysis of Operational Continuity Management in Enterprises

4.1. The Method

A third method involves individual actors and companies that operate critical infrastructures and critical production (National Emergency Supply Agency 2009b). They assess annually their own operational continuity management (ISO/PAS 22399 2007). The method used is capability maturity model (CMM), which originated in the 1980s in Carnegie Mellon University for assessing software processes (Humphrey 1987).

The assessment is subdivided into main headers based on the EFQM Excellence Model (The EFQM):

Management
Personnel
Principles and strategies
Partners and resources
Continuity of operations (general and industry specific)
Measuring results

Each header contains about ten specific questions regarding the manner operational continuity management is performed in the company. Most of the questions are generic and are used for all industries.

The answer to the questions is given as three ticks in boxes: relevant/irrelevant for this company, maturity level at the moment (1–5), and target for next year (1–5).

The levels 1–5 are adapted from CMM for risk management and operational continuity management:

1. Initial	The procedures for continuity management are undocumented, even nonexistent. Incidents are reacted upon in an ad hoc manner. No responsibilities are defined.
2. Repeatable	The key threats and incidents have been identified. The responsibilities have been defined. The personnel know what to do when an incident occurs. The procedures can be repeated, even though written instructions are lacking.
3. Defined	The repeatable procedures are tested and documented. Planning is short term.
4. Managed	The incidents are effectively managed and performance is measured. Planning is long term. The plans and agreements contain alternative ways of acting.
5. Optimised	The documented and managed procedures are continuously improved and audited.

The higher levels build upon the lower levels with added structure in management.

4.2. An Example of Continuity Maturity Level Descriptions in Food Logistics (Figure 6)

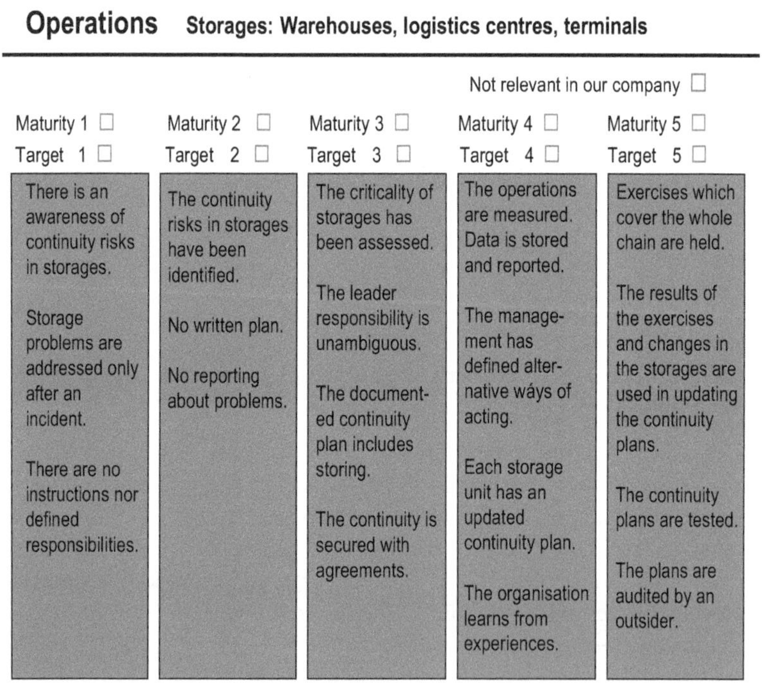

Figure 6. An example of continuity maturity level descriptions (questions) in food logistics

4.3. The Use of the Maturity Assessments

From the national security of supply point of view, the acceptable levels are from level 3 upwards. The question sets have been prepared in the PPP organisation, branch by branch. A maturity assessment tool will be launched as part of a Security of supply extranet portal in 2010 for about 2,000 key companies in Finland.

The maturity assessments of an individual company are confidential and only visible for the company itself. A national, anonymous picture of the maturity in each branch is obtained on a summary level. This summary is used in the industrial committees and in the National Emergency Supply Agency, basically for the same purpose as the Indicators: allocating effort to training, planning, exercising, and perhaps investment or regulation. Each company can benchmark itself against the branch summary.

The maturity descriptions (questions) contain expert knowledge on best practices. In their own way, they promote better operational continuity management. High maturity level descriptions describe good practices, and thus they are guidelines for performance improvement.

Possibly the maturity analysis will replace the indicators, when experience of its usability will be gathered.

5. Appendix – Mathematical Method of the Interdependency Model

5.1. Starting Point

The calculation of relative effects and risks caused by failures is based on expert assessments on three factors pertinent to each infrastructure, basic service, and outside threat (each item in the model) (Sivonen 2005a):

- Dependency
- Mean time between failures
- Duration of failure

5.2. Assumptions

The calculation of effects and risks is based on the following assumptions which simplify the real world:

- The duration of failure can be used as a common measure of adverse effect for all items.
- The causes of failure in one function can be approximated as a linear combination of failures in the whole system of functions.
- The adverse effect of failure grows linearly as the duration of failure grows.
- The occurrence of failure is random. It is independent of previous occurrences.

5.3. Calculation of Effect

The principle of calculating the effect is best understood from an example: Let us assume that in an imagined calculation model there are only three items which depend on each other: electricity, data communication, and payment systems. Let the total adverse effect be 1,000 units (points), if all three items fail for 24 h.[1]

Let our imagined expert assess the dependencies as follows:

- Failures in electricity are 10% of the time caused by failures in data communication, 1% of the time by failures in payment systems (no money, no electricity). The rest, 89% of the time, the reasons for electricity failures are failures in electricity system components, electricity personnel, storms cutting connections, or other things that are not explicitly included in this model. This 89% share will be assigned to electricity itself.
- Failures in data communication are 80% of the time caused by electricity failures and in 1% of the time caused by payment system failures. The rest, 19% of the reasons come from within data communication itself or other sources outside the scope of our simple model.
- Failures in payment systems are 40% of the time caused by data communication failures, 20% of the time by electricity failures, and 40% by others, e.g. data systems.

The first estimate of the indirect effect of an electricity failure is 89% * 333 + 80% * 333 + 20% * 333.[2] = 630. We thus add the shares of electricity in the failures of all three items.

---

[1] The unit could be a monetary unit. The mathematical method would allow for it.
However, loss of money doesn't well describe the effect of failure in hospitals, for example. Therefore, in this study we only calculate total failure time. It will accumulate to those items on which the other items depend most. In this manner we find the relative total effects and risks caused by failures in each item.

[2] In this example the total adverse effect 1,000 has been distributed evenly to the three items, to start with.
So 333 is understood to be the direct adverse effect of 24 h of failure for each of the three items.
It is important to notice, that the final results are independent of this initial distribution.

In the same manner, we calculate the indirect effect of the two other items: data communication 10% * 333 + 19% * 333 + 40% * 333 = 230 and payment systems 1% * 333 + 1% * 333 + 40% * 333 = 140.

Because each item depends on the other two items, this calculation is not sufficient. We should find out the effect of the multistage and circular dependencies. We come closer to the actual total effect by repeating the previous calculation so that we put the results (indirect effects) just obtained in the place of the direct effects of each item (first order effects, 333, where we started from).

Thus electricity: 89% * 630 + 80% * 230 + 20% * 140 = 773, data communication: 10% * 630 + 19% * 230 + 40% * 140 = 163, and payment systems: 1% * 630 + 1% * 230 + 40% * 140 = 65.

It can be proved that by repeating this, the results converge towards numbers that are characteristic to the dependency relations. In this manner the content of the table of 3 * 3 = 9 dependency numbers will be condensed into three total effect numbers. In this case, the total indirect effect of an electricity failure of 24 h is 868, that of a data communication failure is 115 and that of a payment system failure is 16. In this way the original total effect 1,000 of the three items has been redistributed according to the relative total influence of those items.

The convergence can be illustrated and this result obtained by repeating the calculation in a spreadsheet program.

5.4. *Calculation of Risk*

The mean time between failures in each item of the model and in each class of failure duration can be converted into probability numbers of 1, 2, 3, etc. failures per year.[3]

Risk is the expected value total indirect effect of failures (in one unit of time). The risk in one time unit pertinent to one failure duration class and one number (1, 2, 3…) of failure occurrences is obtained by multiplying failure duration by number of occurrences by probability, and by total indirect effect.

Total risk of an item is obtained by adding up the risks pertinent to all numbers of occurrences (1, …, 15 suffice in practice) and all duration classes.

Acknowledgements

A copy of this article will also be published in the web pages of National Emergency Supply Agency www.nesa.fii.

---

[3] The conversion is based on so called Poisson process.

References

The EFQM (2009) (formerly European Foundation for Quality Management) Excellence Model http://ww1.efqm.org/en/Home/aboutEFQM/Ourmodels/TheEFQMExcellenceModel/tabid/170/Default.aspx

Government decision on the targets of security of supply 21.8.2008/539

ISO/PAS 22399 (2007) Societal security — Guideline for incident preparedness and operational continuity management

Humphrey, Watts S. (1987) Characterizing the Software Process: A Maturity Framework. IEEE Software, vol. 56, no. 2, March 1988, pp. 73–79.

Laki huoltovarmuuden turvaamisesta 18.12.1992/1390 (Security of Supply Act) http://www.finlex.fi/fi/laki/ajantasa/1992/19921390

Mathews, John H. (2004a) The Power Method for Eigenvectors. California State University Fullerton, Department of Mathematics. http://math.fullerton.edu/mathews/n2003/PowerMethodMod.html

Mathews, John H. (2004b) Power Method Proof. California State University Fullerton, Department of Mathematics. http://mathews.ecs.fullerton.edu/n2003/powermethod/PowerMethodProof.pdf

National Emergency Supply Agency (2009a) Internet pages http://www.huoltovarmuus.fi/ http://www.nesa.fi/

National Emergency Supply Agency (2009b) System definition for the new extranet HUOVI

Nenadic, Zoran (2002) Counting Process. California Institute of Technology, Department of Mechanical Engineering. http://robotics.caltech.edu/~zoran/Research/poisson/node1.html

Sivonen, Hannu (2005a) Yhteiskunnan huoltovarmuuden kannalta keskeisten toimintojen riskiarviointi (Finnish for Risk Assessment of Functions Vital to Society's Security of Supply) Publications of National Emergency Supply Agency # 2/2005 http://www.huoltovarmuus.fi/documents/3/HVK_JULK_2005 2_Keskeisten_toimintojen_riskiarviointi_web.pdf

Sivonen, Hannu (2005b) Internal instructions for the National Emergency Supply Organisation on security of supply indicators

Food Chain *Defense* and Its Potential Implications on Traditional Foods: The Portuguese Case

Alexandra Veiga[1] **and José Empis**[2]

[1] Instituto de Tecnologia Química e Biológica, Universidade Nova de Lisboa, Av. da República, 2780-157 Oeiras, Portugal; e-mail: aveiga@itqb.unl.pt

[2] Institute for Biotechnology and Bioengineering, CEQB do IST, Av. Rovisco Pais, 1049-001 Lisboa, e-mail: jempis@ist.utl.pt

Abstract

Traditional foods are presented as important components of local diets characterised by unique dispersed production, hence providing specific scenarios of compliance with Food Safety and Food *Defense* practices. The evolution of attitudes towards Food Safety and Food *Defense* during the last years, with the concomitant and resulting legislative changes, has affected traditional foods. Their present and future consequences for the production and consumption of these important cultural, societal and economical components of local diets are discussed, and the Portuguese case is presented in more detail.

1. Introduction

Some recent historical events have awakened European Governments to the vulnerability of the food chain, and its characteristics as a potential target for a terrorist attack. In first place the food crisis in the 1990s, with the outbreak of the Mad Cow Disease in the UK and other European countries and the dioxin crisis in Belgium, shook consumers' confidence in food and highlighted the vulnerability of this system. Secondly, the events of September 11, 2001 followed by bioterrorist menaces consisting mainly of mailed envelopes containing white dust, allegedly with spores of the potentially lethal bacterium *Bacillus anthracis*, brought to mind that many vectors, food included, could theoretically be used to deliver a pathogenic agent (biological, chemical or radioactive).

As a result of the European food crisis, a White Paper on Food Safety was published by the European Commission in 2000 (European Commission 2000) and the General Food Law – Regulation EC/178/2002 (Regulation (EC) 2002) – was published setting up the European Food Safety Authority (EFSA) as an

H. Alpas and B. Cirakoglu (eds.), *Food Chain Security*,
DOI 10.1007/978-90-481-9558-9_10, © Springer Science + Business Media B.V. 2010

organization in charge of performing an independent food (and feed) risk assessment and communication. Further, the scope and procedures of the Rapid Alert System for Food and Feed (RASFF), a system which had been in place since 1979 and that provides the control authorities with an effective tool for exchange of information on measures taken to ensure food safety, were redefined by this Regulation.

As a consequence of all these changes in the food safety approach in the European Union, some Member-States felt the need to re-organize their food Risk Analysis structures (Risk Management, Risk Assessment and Risk Communication) in order to make them able to fit the requirements of the Commission and to respond to more demanding consumers, as well as to efficiently act as a contact point for EFSA.

The re-organization of these structures in charge of guaranteeing food safety, together with measures taken in relation to preparedness towards terrorist attacks, are key steps in the process of building up a proactive approach regarding Food Chain *Defense*.

The question is how far the introduction of such deep changes in food production systems will affect small scale manufacturers that contribute a large share of traditional foods production. And whether these products will preserve their original characteristics upon compulsory changes resulting from the novel demands – legal, economic or of another sort – aiming at guaranteeing stronger food safety standards as well as protection from intentional food contamination.

Portugal is a European Union (EU) Member-State with strong gastronomic traditions, and foods with unique characteristics that are an important part of its cultural patrimony. The Portuguese experience resulting from reorganization of food safety structures in order to fit EU and consumers' demands, bearing in mind that attention should be paid to food *defense*, as well as its putative impacts on traditional foods, will be used to discuss these issues.

2. Food Safety in Portugal in the Last 2 Decades

By the turn of the century, the Portuguese situation regarding food risk analysis could be considered complex, heavy and confusing. In the 1990s a multiplicity of state departments belonging to at least three ministries (Ministry of Agriculture, Ministry of Health and Ministry of Economy) was involved in risk analysis activities. In this context the Ministry of Agriculture assumed the greatest visibility with several Departments carrying out activities of risk management (food control and regulation), risk communication and risk assessment. Ministries of Health and Economy, mainly the last one, had also an important share in the process. That meant the existence of many gaps and overlaps of functions and conflicts of competences, which strongly came into light during food crisis such as the mad cow disease that highly affected Portugal, but also in minor crisis. The general approach was reactive either than proactive, with no clear strategy for risk assessment and risk communication.

Awareness of the ineffectiveness of the existing system led to the recognition of the urgent need for its re-organization. However, between 2001 and 2004 many unsuccessful attempts were made to create a Food Safety Agency in which the two main different models considered: risk management separated from risk assessment and risk communication, and these three components of risk analysis put all together. By end 2004, the Portuguese Food Safety Agency (APSA) was eventually set up as the organization in charge of food risk assessment and risk communication, but the status quo from the 1990s in food control and other components of risk management was maintained.

The turning point only came January 1, 2006, with the creation of the Economy and Food Safety Standards Authority (ASAE) (Decreto-Lei nº 237/2005 de 30 de Dezembro de 2005). It merged at least eight directorates formerly from the Ministries of Agriculture and Economy, including the recently created APSA. ASAE is competent in a wide area of economic activities and is also the Authority responsible for Food Safety, including Feed, having a total staff of over 600 workers, covering several professional skills. In the Food Safety area, ASAE develops its activities within the three main components of Risk Analysis, with the exception of Food Regulation. Concerning risk management, ASAE has law enforcement as well as police and crime investigation powers. Risk assessment and risk communication are dealt within a Scientific Directorate that is also responsible for the coordination of all the activities related to EFSA including the Portuguese representation in EFSA's Advisory Forum.

The organic structure of this Authority covers the whole country and several Central and Regional Control Services (Decreto-Lei nº 274/2007 de 30 de Julho de 2007).

The "arrival" of ASAE with the adoption of a proactive instead of a reactive behaviour created a new dynamic way of using the available resources benefiting from careful strategic planning.

3. Traditional Foods

Traditional foods are important components of diets everywhere, and normally linked to specific countries, or even regions. Traditional foods are sometimes thought to be mainly produced in small quantities in a diversity of locations, especially 'at home', but may be commercialized by myriad small producers as well as by industrial concerns. One can thus find within traditional foods a diversity of situations regarding the observance of food safety principles. Increased biodiversity, as is to be expected in warmer regions of the world, may help increase the number of traditional foods which are locally produced and consumed. This is clearly the case of the Mediterranean countries, in which Portugal is also definitely included despite the fact that this sea does not bathe its coastline, because of the similarity of climate and produce (both from agriculture and the sea).

Traditional food existence and diversity in a given region is also affected by historical factors. One good example is the existence of a traditional charcuterie produced from the Iberian Pata Negra pig. These products depend on the local availability of *Quercus ilex* and maybe also of *Quercus suber* glands, as well as of the genetics of this *Cerdus ibericus* pig, because they are the result of a special feeding system based on the glands of those trees, by a specialised breed. For many centuries, nevertheless, the whole of Iberia (in fact all the way up north to Poitiers), was a Muslim area, depending from Granada. It speaks strongly of their regime's tolerance that despite considering pork unclean, the possession and use of pork products by the local populations was tolerated, maybe somewhat unofficially, but ensured the continuation of a traditional means of livelihood under the local climate and soil conditions, as well as the sustainable maintenance of the *Quercus* forest.

Portugal is the European country where the *per capita* consumption of soup is larger, mainly vegetable soup. Indeed, only in Thailand and China does soup represent a more important part of the daily diet than in Portugal, and in Europe the second highest *per capita* soup consumption is France, but at less than half the value it has in Portugal (personal communication from an element of McDonald's staff in Portugal). This probably has a historical reason and represents an influence of Oriental diets which the Portuguese brought back from their seafaring expeditions. One of its consequences is that McDonald's decided to add soup to its menu in Portugal. Soup must thus be considered together with coriander leaf, typically used around the Pacific rim and brought by land to the Arab world, which is cultivated in Portugal and included in many traditional food recipes. Portuguese soup traditionally uses various *Brassica* species (among which *Couve Galega* or cow cabbage, *Couve Repolho* or drumhead cabbage, couve Portuguesa or Portuguese kale and *Couve Coração de Boi* which is another kind of drumhead cabbage are perhaps the most important), some of which strongly regional, and is because of that a functional item in Portuguese food. But probably the most documented component of Portuguese food is codfish, *Gadus morhua*, which is captured in the Northern Atlantic, from which Portuguese ships traditionally caught it since the late fifteenth century, its population having nowadays dwindled badly due to the joint effects of overfishing, bottom trawling as well as a temporary increase of the grey seal (*Halichoerus grypus*) population may all have had a role in this strong decrease of cod population (http://www.hsicanada.ca/pdfs/fast_facts_grey_seal_hunt.pdf). Cod was traditionally processed by salting water out and sun-drying the salted fish when boats got home from the Canadian coast. From the period of Arab dominance marinated fish has been included in the Portuguese diet, consisting essentially of previously fried fish preserved in vinegar, onion and oil added. Last but by no means least one should consider desserts. Portuguese desserts are unequivocally linked with convent life. They consist mainly of eggs, almond, and sugar. Less important ingredients are fruits, especially figs and Malabar squash. Cane sugar became widely available during the sixteenth century, and the Malabar squash too. Sugar was initially an extremely expensive ingredient, but the organisation of its production mainly in Central and South America lead historically to an eventual surplus and it tended to replace formerly used honey as a sweetener.

Dairies are an important component of the Portuguese diet, and among them cheese and cottage cheese. Traditional cheeses in the Azores are produced with cow milk, but in the mainland ewe milk is dominant, and goat milk is also present. The consumption of cream cheese and cottage cheese is traditional, and the milk used in their production is pasteurised. Most traditional cheeses are nevertheless produced from non-pasteurised milk, and thus require a maturation period of at least 60 days prior to consumption, in order to ensure the elimination of most potentially pathogenic microorganisms. In many cases shorter maturation periods (22 days) are used, but in those cases a strict surveillance of the animals and facilities is mandatory (because of brucellosis and tuberculosis). *Listeria monocytogenes* is also a strong preoccupation in these products, and is hence under surveillance.

One should probably recall that EU law permits food irradiation (Directive 1999/2/EC; Directive 1999/3/EC) when this is technologically recommended. The similarity of the traditional Portuguese cheeses with short maturation periods and the French camembert cheeses would recommend, eventually, that such an irradiation be conducted, though there are no adequate premises in Portugal for that purpose.

Charcuterie is traditional Portuguese food even when it is not produced from the Pata Negra pig. In fact some of the Portuguese traditional sausages may even contain meat from other animals. This is especially the case of *alheiras*, allegedly invented by crypto-Jewish people after Jews had been evicted from Portugal or converted, which contained, besides pork, chicken, garlic, paprika, olive oil and salt, as well as bread. As most Portuguese traditional sausages, its preservation process includes smoking. Other important procedures for traditional sausage preservation include the use of nitrates/nitrites, cooking, and lactic acid fermentation. Most of these products are consumed after cooking, often frying procedures, which ensures a high-temperature stage useful in diminishing the population of potential pathogens such as *Salmonella* spp., *Listeria monocytogenes*, *Campylobacter jejuni*, as well as the parasite *Trichinella*, among others capable of inducing dangerous cases of food poisoning. The absence of these and other pathogens from the traditional sausage product which are to be consumed without further cooking makes it mandatory that hygienic processing be used in their manufacture, topped by close control measures.

As may be seen in the description above, traditional Portuguese food is either osmotically preserved with salt or sugar, or has suffered a cooking procedure that strongly diminishes existing microbial contaminants. It thus presents low intrinsic potential for food-borne poisoning, and simple HACCP procedures are thought to adequately protect consumer's interests. The main problem arises due to a local tendency for the multiplication of micro-producers, including private homelike kitchens, where HACCP tends not even to have been heard of. Disciplining this situation has proven a reasonable challenge, because of the consequences it carried over small parallel economy transactions, but seems to be largely under way.

Traditional foods have presented some of the most delicate problems for European citizens during the process of necessary uniform legalization. The most cited of these problems related to the concept linking starting materials and/or processes to end product denominations, as e.g. was the case for beer in Germany (http://aei.pitt.edu/5643).

The European Food Information Resource Network (EuroFIR) has as an aim to provide comparable or harmonized data on the nutritional composition of traditional foods across selected European countries by chemically analysing selected recipes and harmonising existing compositional data (Trichopoulou et al. 2007). Some examples of Portuguese traditional recipes to be included in EuroFIR pilot study are green cabbage soup (*caldo verde*), Portuguese boiled dinner (*cozido à portuguesa*), or egg sweet from Murça (*toucinho do céu de Murça*), and other diverse recipes from other countries like potato dumplings (Austria), waffles from Brussels (Belgium), rose jam (Bulgaria), sauerkraut (Germany), olive bread (Greece), pickled blood pudding (Iceland), tree cake (Poland), Galician octopus (Spain), or baklava (Turkey) (Trichopoulou et al. 2007).

4. Will Traditional Foods Easily Survive Increasingly Strict Food Safety Systems?

The activity of the Portuguese food safety authority started exactly the same day as Regulation (EC) Nº 852/2004 of the European Parliament and of the Council of 29 April 2004, on the Hygiene of foodstuffs (Regulation (EC) No 852/2004), became valid. This Regulation imposes food business operators enact procedures based on the HACCP principles. This meant that a more demanding legislation coincided with the introduction of deep changes in the establishment, since the new authority adopted a pro-active rather than a reactive behaviour, carrying out an intense inspective activity, breaking down the general feeling of impunity that was perceived among stakeholders.

The activity of ASAE was intensively followed by the media and in just a couple of months it became extremely notorious gaining wide media coverage. As a result, a strong movement of public opinion on food safety national and European policies was triggered. In spite of the general recognition that a strong food safety authority and clear food laws were important to guaranty consumers' health protection, voices raised against what, in their view, were too heavy laws and too strict interpretations, compromising the survival of some small economic agents that wouldn't be able to cope with them.

These reactions were shared by many different sectors of the society: economic agents and their associations, consumers, "opinion makers" and the media in general. They were particularly concerned with traditional foods, the survival of some of which could eventually be in danger if these policies would continue to be followed.

This kind of reaction was not exclusive to Portugal but also happened in other countries with strong gastronomic traditions or even in others where gastronomy is not that important.

Although as regards food *defense*, traditional foods present little differences from other foods, because intentional alteration of their quality and especially their safety is similar, a health problem associated with a National traditional product could seriously affect a country's reputation, and have a strong economic impact.

An important question is "what is a traditional food"? According to the Commission Regulation (EC) nº 2074/2005 of 5 December (2005), foods with traditional characteristics' means that, in the Member State in which they are traditionally manufactured, are: (i) recognised historically as traditional products, (ii) manufactured according to codified or registered technical references to the traditional process, or according to traditional production methods, or (iii) protected as traditional food products by a Community, national, regional or local law. Although it is common that people have the perception of what is a traditional product in their own country or even in a foreign country, very often they are mistaken and the border between traditional and non-traditional foods may not be easy to establish.

In the past, traditional food was produced in small amounts and sold in markets near the production region. Nowadays, this food can be produced either in small or large amounts, and it can be sold near the production region or travel around the world to be commercialized very far from its origin. This brings additional problems. When food travels, there are more opportunities for potential hazards to show up and, in the case of those from biological origin, to multiply. This means that if food travels, hazards may also travel. Some examples of pathogens that are likely to cause problems in traditional foods are *Listeria monocytogenes* (cheese and sausages) (for e.g. Olivier et al. 2005; Ferreira et al. 2007), *Salmonella* (almost anywhere, but particularly in desserts manufactured with raw eggs), *Brucella* (cheeses manufactured with raw milk) (Memish and Balkhy 2004) and *Escherichia coli* and *Staphylococcus aureus* (associated with deficient hygiene practices).

In any case, traditional foods will probably not represent a major risk for malicious attack due to the limited diffusion which they possess. The relatively large number of sources originating them and the relatively low volumes or percentages ensured by any individual producer ensure this characteristic and may probably discourage malicious attacks except those motivated against specific producers.

4.1. European Legal Framework

In order to protect traditional foods, European legislation has been set up. Some of its aspects that are considered to be the most relevant in the present context will be mentioned.

The General Food Law 178/2002 (Regulation (EC) No 178/2002), that provides the basis for the assurance of a high level of protection of human health and consumers' interest in relation to food, states that account should be taken in particular of the diversity in the supply of food including traditional products. The above mentioned Regulation 852/2004 on the hygiene of the foodstuffs also refers that "flexibility is (...) appropriate to enable the continued use of traditional methods at any of the stages of production, processing or distribution of food and in relation to structural requirements for establishments. Flexibility is particularly important for regions that are subject to special geographical constraints (...). However, flexibility should not compromise food hygiene objectives. Moreover, since all food produced in accordance with the hygiene rules will be in free circulation throughout the Community, the procedure allowing Member States to exercise flexibility should be fully transparent (…).

This same regulation states that Member States may adopt national measures adapting the requirements of this law with the aim to enable the continued use of traditional methods. However, this is a process that involves huge bureaucratic steps requiring, for example, the description of each foodstuff and establishment concerned. In countries like Portugal where there are inumerous traditional foodstuffs produced by many different producers it will involve an endless task.

Finally, Regulation 2074/2005 (Commission Regulation (EC) No 2074/2005) states, once more, that flexibility is needed so foods with traditional characteristics can continue to be produced, and that food with traditional characteristics should therefore be defined and general conditions applicable to such foods should be laid down. It also allows Member States to grant establishments manufacturing foods with traditional characteristics individual or general derogations.

In summary, EU legislation applies to foodstuffs in general, traditional foods included, but it is possible to apply derogations (exceptions of a temporary nature) to traditional foodstuffs/producers if food hygiene should not be compromised. However, putting in place such derogation is not easy. There is a heavy bureaucratic procedure to obtain derogations, some countries have a huge number of traditional products, application has to be made for each product/producer, and knowledge on some traditional products is confined to a small number of illiterate people (very low capacity to apply for derogations). As a consequence, some products may not survive. On the other side, food safety should not be compromised, since it is about consumers' health that it stands for. It is therefore very difficult to find the right equilibrium.

4.2. Looking at Traditional Foods in Europe

Vulnerability of traditional foods is a subject that has been worrying many countries around Europe, and EU itself. These concerns were recently expressed by European Commission's Directorate-General for Health and Consumers on the sidelines of the dialogue in a meeting in Malaysia on the EU-Asian Cooperation on Codex Matters. He stated that he believed that traditional food is not necessarily unsafe to eat even though some of the ingredients used in its preparation are considered unhealthy. He referred the belief of the European Commission that the *Codex Alimentarius* Commission or Codex should take into account traditional food and protect the food that has become typical in people's lives. Codex, established by the United Nation's Food and Agriculture Organisation and the World Health Organisation, is the body responsible for developing food safety standards. The commissioner referred that many traditional foods had lost their authenticity because of the change in the way they were prepared either to make them healthier or that some of the ingredients, such as agricultural products, no longer had the authentic taste because of the effects of climate change or chemicals used to grow them. However, it is a wish of the EU to keep this food as it is because there is a long traditional story which has kept the people together. He affirmed that EU was trying to push the Codex to look into protecting traditional foods which had become the staple diet of many people around the world. This could include hygienic food preparation, safe and healthy ways of managing crops as well as providing guidelines on daily intakes of traditional food, he added (Daily Express, Malaysia, 23 February 2009).

Many initiatives have been undertaken regarding traditional foods and their future in Europe and all over the world. Some examples can be given. A research project, TRUEFOOD – Traditional United Europe Food, started in 2006 with the support of the Commission. The overall aim of TRUEFOOD is to introduce suitable innovations into traditional food industry to maintain and increase the competitiveness of the industry in an increasingly global European market place (http://www.truefood.eu). A conference in the framework of the French Presidency of the European Union Council on TRADITIONAL FOODSTUFFS – "Food in Europe: Diversity and Safety" was held in Paris on October 2008 (http: //www.ue2008.fr/impressionPDF6f59.pdf?url=%2FPFUE%2Flang%2Fen%2Faccueil %2FPFUE-10_2008%2FPFUE-23.10.2008%2Fproduits_traditionnels_et_securite_ des_aliments). An update on food hygiene legislation was also made in order to protect small food business operators: an ammendment to Article 5 (1) of Regulation 852/2004 that requires all food business operators to install, implement and maintain a permanent procedure or procedures based on HACCP principles was produced to exclude from the whole of the Article 5 (1) requirement every food business operator with fewer than 10 employees who predominantly sell food direct to the final consumer. Excluded food business operators would still have to comply with all other relevant requirements of Regulation 852/2004 (Regulation

(EC) No 852/2004). In the framework of the International Polar Year – an activity was proposed entitled "Integrity of the Traditional Food System and Environmental Health in the Circumpolar North" (http://classic.ipy.org/development/eoi/proposal-details.php?id=384). The above mentioned EuroFIR pilot study is itself an outcome of a concern to protect traditional foods. The importance of traditional foods will also probably motivate researchers to devise novel and more performing analytical systems, both of chemical and biological nature, to characterise their authenticity as well as their innocuity.

4.3. Traditional Foods and Food Defense

What if food *defense* issues are added to traditional food production? Apart from measures aiming at preventing food from accidental hazards (HACCP would guarantee it), food business operators would have to take measures to prevent intentional hazards. These include the security of the premises, the surveillance and monitoring of activities so that identification and prevention of acts intended to disrupt food supply can be implemented, personnel security, and emergency responses. This means that if these practices would be adopted, production of traditional foods would be even more difficult, requiring additional skills and, in some cases, changes in facilities, resulting in an increase in the costs associated with production. These kinds of measures are usually not compulsory and, most probably, they will not be in the future. But pressures to implement them might come from insurance companies, in order to decrease the insurance premiums. This factor is presently particular important in multinational companies which, in general, don't produce traditional products. But this might change.

If one handles traditional foods in a more relaxed system, opening exemptions in legal requirements, they can become an attractive target for a terrorist attack. And they will be really attractive if the aim would be to disturb a country's or a region's image.

One does not know what the future will be, but if one would increase the requirements of a food safety system, food *defense* issued included, this would add difficulties in obtaining derogations for certain food products. This would mean that production would require higher qualified skills from the personnel and would be more expensive. As a result, some traditional foods may really be threatened.

On the other hand, if more exemptions are to be opened for traditional foods, the control system would be less demanding with no security measures, which, ultimately, would result on the survival of more traditional foods, though at risk of an increased vulnerability to a malicious attack.

However, successful cases where traditional food was produced under the best practices to guarantee food safety are reported. An interesting case was the application of GMP and HACCP to traditional food processing at a semi-commercial Kenkey production in Ghana (Kenkey is a staple dish similar to a sourdough dumpling from the Akan, Ga and Ewe inhabited regions of West Africa, usually served with a soup, stew, or sauce (http://en.wikipedia.org/wiki/Kenkey). This application proved as effective as a quality management system (Amoa-Awua et al. 2007).

5. The Portuguese Case

5.1. Regarding Traditional Food

In Portugal, food is very important for social, cultural and economical reasons. Like in many other countries, every social occasion happens around a table and recipes are passed throughout generations. Some traditional products became a label of certain geographical regions, and can represent an important share in that region's income. Moreover, tourism represents 11% of total national income and 10% of employment (official figures, available at http://www.planotecnologico.pt/document/Doc_12.pdf), and traditional food plays an important role in it.

The discussion on traditional food's survival, resulting from the changes mentioned above, has triggered some initiatives from different stakeholders.

Two political parties have presented to the Parliament projects aiming to recommend the Government prepare specific measures to protect traditional products and practices (http://jpn.icicom.up.pt/2008/03/05/parlamento_discutiu_regime_especial_para_produtos_tradicionais.html).

As result of the initiative of the political party in charge of Government in January 2008, a working group on small producers/ traditional products was set up by the Commission of Economical Issues, Innovation and Regional Development of the Parliament. The aim of this working group was to address the most relevant questions related to the production and protection of traditional products and the necessity to keep assuring food safety. A report was published in July 2008 (Comissão de Assuntos Económicos 2008) and the major outcomes were the following proposals: to set up a more effective legal and regulating system that allows small producers or its representatives to propose derogations; to create a simplified information system that allows citizens to access legal obligations regarding food safety (in particular it is suggested that documents to be published on the Internet on "what to do" on the production of traditional foods should be produced); the revision and clarification of criteria used in the application of the word "artisanal" on food products; to demand the official entities involved in intellectual property registration and in food and labelling inspection to guarantee that "Protected Designation of Origin" (PDO) and "Protected Geographical Indication" (PGI) denominations are respected; to recommend that the Government promotes the creation of information posts advertising local products, food included, in service stations in motorways, and to develop partnerships between

Central Administration and municipalities in order to make available services to inform, give training and advise for activities related to traditional products (Comissão de Assuntos Económicos 2008).

One month later, changes in law were produced with the aim to protect traditional products, simplifying and facilitating the activity of small producers and homemade productions (Despacho Normativo nº 38/2008).

An initiative coming from the municipalities and producers, resulting from awareness of how important traditional foods were to their region, ended up in the constitution of an association in May 2008 with the aim to protect these products (QUALIFICA – National Association of Municipalities and Producers in order to Valorise and Qualify Portuguese Traditional Products (http://www.qualifica.pt).

5.2. Regarding Food Defense

In Portugal there is no record of a single malicious attack to the food supply chain. However, there have been some cases of blackmail or sabotage, where no political or religious motivations exist but only personal interests. Some episodes involve workers in the food industry that deliberately contaminate a product just because they were unhappy with their boss or have personal problems. Others are performed by consumers that state that they have contaminated a product, with lye for example, without identifying it and claiming for money to give the information. There have also been cases where consumers threaten to contaminate a product if the company doesn't give them something they want. Curious cases of people asking for scooters or trips have been registered. None of these cases have had serious consequences for consumers or producers. However, all the players that have a role in the food chain should be prepared to prevent terrorist attacks, or to minimize its impact where avoidance was not possible.

Following the events of September 11, a Contingency Plan for Health was set up in 2002, aiming at minimising the possible consequences of a deliberate biological incident. Its objectives were to guarantee the rapid detection of potential biological agents and/or of cases of disease resultant from their deliberate release; to define measures for appropriate treatment and prophylaxis; and to guaranty the restriction of the dissemination of biological agents and/or cases of disease. This is a comprehensive plan mainly addressed to the National Health Service's professionals and structures, but to be carried out in collaboration with other Ministries and to take in account the Emergency Plan from the Civil Protection. Due to its nature, it does not pay particular attention to the food chain or to chemical or radiological agents.

Some years later, and coinciding in time with the creation of ASAE, the Portuguese Army inaugurated the twin laboratories – the Laboratory of Bromatology and the Laboratory of Biological *Defense*, that were set up to analyse biological factors related to terrorist events. Together, they aim to study, identify and neutralise bacteria, modified bacterial strains or other microorganisms with potential to be used in bioterrorist attacks, and to act in the identification of biological pathogens that may (accidentally or deliberately) contaminate food.

Since the Portuguese Contingency Plan for Health did not pay particular attention to the food chain, setting up of ASAE and the Laboratories from the Army created conditions to fulfil this gap.

Participation of these two organizations in a meeting of the NATO Pilot Study on the Food Chain Security (http://www.nato.int/science/pilot-studies/fcs/fcs-index.htm) in Lisbon, was very important to gather them into the context of food *defense*, and contributed to increase the awareness that there was an urgent need for the main intervenients to increment their capacities of prevention and response to a deliberate contamination of the food supply. As an outcome, a working group on food terrorism (GTTA) was set up in 2007. This working group is coordinated by ASAE and includes members of the Laboratories from the Army and of the National Health Institute (INSA) that has been strongly involved in the working up of the Contingency Plan for Health. GTTA's mission is to provide guidance for food producers, retailers and other relevant economic agents in a case of a potential terrorist attack, and to elaborate a contingency plan specific to the food sector. Its main objectives are: to produce a document for food industry, retailers and other relevant economic agents in order to provide guidance for preparedness to a potential food terrorist attack (Guidelines); to give support to these agents on the application of these Guidelines; to develop a Contingency Plan for a Situation of a Terrorist Attack to the Food Supply Chain; to articulate the above mentioned Contingency Plan with the Portuguese Contingency Plan for Health. Some general strategies were defined for pursuing said objectives. In particular, recommendations of the World Health Organization (WHO) to strengthen prevention and response in national systems will be adopted (WHO 2002); the documents will be prepared in close relationship with economic agents from all the food chain sectors (agriculture, food processing, storage and transportation, retailers, and catering) and their respective professional associations; an agreement platform with the economic agents and their respective professional associations will be established; the *Carver Plus Shock Method* developed by the Food and Drug Administration (FDA) will be included in the recommendation to the economic agents; and the most probable scenarios of food chain attack will be built up.

5.3. The Future

Some positive measures were already taken regarding preservation of Portuguese traditional food, but much additional work is yet to be done. Changes of living patterns and eating habits make the task of preserving some foods difficult, since knowledge on their preparation and authenticity is being progressively lost. It is crucial to prepare an exhaustive list of original traditional products and recipes, so that measures for their preservation can be effectively applied. Demand for specific derogations could be very important where traditional ways of preparation, although not always complying with legal requirements, are determinant in the originality of the end product.

However, food safety should never be compromised since it aims at assuring consumers' health. Professional associations could play a determinant role in preserving traditional foods, organising training sessions on food safety addressed to those small producers that have the knowledge on how to produce them. And also training sessions on the original methods of production, for those who look at them as an interesting and promising growing market.

It will be a challenge to include food *defense* issues in this context, since it is not an easy task at the present time. But interested players should start preparing the foundations for this scenario, since it is very likely that this will be the one future will bring.

6. Final Remarks

According to a statement of members of a United Nations agency, dated September 2009, traditional foods are threatened by the spread of Western eating habits through globalisation (http://www.france24.com/en/20090908-globalisation-threatens-indigenous-foods-un-agency). If specific attention is not paid to these products, many of them will disappear, with immeasurable losses in certain regions' culture, identity and economy. Governments should have this in mind, providing tools that stimulate their production, without forgetting that excessively permissible measures aiming at counteracting this tendency could make these products more vulnerable to a food chain attack.

The Portuguese case presented herein, constitutes a reflexion on how difficult and sensitive the issue of traditional foods is, and on the challenges they have to face to be able to survive this demanding but vulnerable globalised world.

References

European Commission (2000) 'White Paper on Food Safety in the European Union' COM(99)719, 12 January 2000; available at http://ec.europa.eu/dgs/health_consumer/library/pub/pub06_en.pdf (accessed on September 10, 2009)

Regulation (EC) No 178/2002 of the European Parliament and of the Council of 28 January 2002 available at http://eur-lex.europa.eu/pri/en/oj/dat/2002/l_031/l_03120020201en00010024.pdf (accessed on September 10, 2009)

Decreto-Lei nº 237/2005 de 30 de Dezembro de 2005

Decreto-Lei nº 274/2007 de 30 de Julho de 2007

http://www.hsicanada.ca/pdfs/fast_facts_grey_seal_hunt.pdf (accessed on September 10, 2009)

Directive 1999/2/EC of the European Parliament and of the Council of 22 February 1999 on the approximation of the laws of the Member States concerning foods and food ingredients treated with ionising radiation, Official Journal L 066, 13/03/1999 P. 0016–0023 (available at, http://eur-lex.europa.eu/LexUriServ/LexUriServ.do?uri=CELEX:31999L0002:EN:HTML)

Directive 1999/3/EC of the European Parliament and of the Council of 22 February 1999 (available at http://eur-lex.europa.eu/LexUriServ/LexUriServ.do?uri=OJ:L:1999:066:0024:0025:EN:PDF)

http://aei.pitt.edu/5643 (accessed on September 10, 2009)

Trichopoulou A, Soukara S and Vassilopoulou E (2007) Traditional foods: a science and society perspective. Trends Food Sci Technol 18: 420–427.

Regulation (EC) No 852/2004 of the European Parliament and of the Council of 29 April 2004.

Commission Regulation (EC) No 2074/2005 of 5 December 2005 laying down implementing measures for certain products under Regulation (EC) No 853/2004 of the European Parliament and of the Council and for the organisation of official controls under Regulation (EC) No 854/2004 of the European Parliament and of the Council and Regulation (EC) No 882/2004 of the European Parliament and of the Council, derogating from Regulation (EC) No 852/2004 of the European Parliament and of the Council and amending Regulations (EC) No 853/2004 and (EC) No 854/2004 (available at http://eur-lex.europa.eu/LexUriServ/LexUriServ.do?uri=OJ:L:2005:338:0027:0059:EN:PDF)

Olivier SP, Jayarao BM, Almeida RA (2005) Foodborne pathogens in milk and the dairy farm environment: food safety and public health implications. Foodborne Pathog Dis 2: 115–129.

Ferreira V, Barbosa J, Silva J, Felício MT, Mena C, Hogg T, Gibbs P, and Teixeira P (2007) Characterisation of *alheiras*, traditional sausages produced in the North of Portugal, with respect to their microbiological safety. Food Control 18: 436–440.

Memish ZA and Balkhy HH (2004) Brucellosis and international travel. J Travel Med 11: 49–55.

Daily Express, Malaysia, 23 February 2009 (available at http://www.dailyexpress.com.my/print.cfm?NewsID=63065, accessed on September 17, 2009)

http://www.truefood.eu

http://www.ue2008.fr/impressionPDF6f59.pdf?url=%2FPFUE%2Flang%2Fen%2Faccueil%2FPFUE-10_2008%2FPFUE-23.10.2008%2Fproduits_traditionnels_et_securite_des_aliments (accessed on September 9, 2009)

http://classic.ipy.org/development/eoi/proposal-details.php?id=384, accessed on September 11, 2009

http://en.wikipedia.org/wiki/Kenkey

Amoa-Awua WK, Ngunjiri P, Anlobe J, Kpodo K, Halm M, Hayford AE, and Jakobsen M (2007) The effect of applying GMP and HACCP to traditional food processing at a semi-commercial kenkey production plant in Ghana. Food Control 18: 1449–1457.

http://www.planotecnologico.pt/document/Doc_12.pdf (accessed on September 17, 2009)

http://jpn.icicom.up.pt/2008/03/05/parlamento_discutiu_regime_especial_para_produtos_tradicionais.html

Comissão de Assuntos Económicos, Inovação e Desenvolvimento Regional, Grupo de Trabalho dos Pequenos Produtores/Produtos Tradicionais. Assembleia da República. Relatório. Lisboa, 14 de Julho de 2008.

Despacho Normativo nº 38/2008 do Gabinete do Ministro da Agricultura do Desenvolvimento Rural e das Pescas, de 13 de Agosto de 2008.

http://www.qualifica.pt

http://www.nato.int/science/pilot-studies/fcs/fcs-index.htm

WHO (2002) "Terrorist Threats to Food – Guidance for Establishing and Strengthening Prevention and Response Systems".

http://www.france24.com/en/20090908-globalisation-threatens-indigenous-foods-un-agency

Subject Index

FSC
www.fsc.org
MIX
Papier | Fördert
gute Waldnutzung
FSC® C083411

Kolibri 360 GmbH
Ferdinand-Jühlke-Straße 7
99095 Erfurt, Deutschland
produktsicherheit@kolibri360.de